**INVESTIGATIONS IN NUMBER, DATA, AND SPACE®**

# The Number System

# Building Number Sense

## Grade 1

Marlene Kliman
Susan Jo Russell

*Developed at TERC, Cambridge, Massachusetts*

**Dale Seymour Publications®**

**White Plains, New York**

The *Investigations* curriculum was developed at TERC (formerly Technical Education Research Centers) in collaboration with Kent State University and the State University of New York at Buffalo. The work was supported in part by National Science Foundation Grant No. ESI-9050210. TERC is a nonprofit company working to improve mathematics and science education. TERC is located at 2067 Massachusetts Avenue, Cambridge, MA 02140.

This project was supported, in part, by the
**National Science Foundation**
Opinions expressed are those of the authors and not necessarily those of the Foundation

This book is published by Dale Seymour Publications®, an imprint of Addison Wesley Longman, Inc.

Dale Seymour Publications
10 Bank Street
White Plains, NY  10602
Customer Service: 800-872-1100

Managing Editor: Catherine Anderson
Series Editor: Beverly Cory
ESL Consultant: Nancy Sokol Green
Production/Manufacturing Director: Janet Yearian
Production/Manufacturing Manager: Karen Edmonds
Production/Manufacturing Coordinator: Shannon Miller
Design Manager: Jeff Kelly
Design: Don Taka
Composition: Andrea Reider
Illustrations: DJ Simison, Rachel Gage, Carl Yoshihara
Cover: Bay Graphics

 Printed on Recycled Paper

Order number DS43703
ISBN 1-57232- 467-8
6 7 8 9 10-ML-01 00

# INVESTIGATIONS IN NUMBER, DATA, AND SPACE®

T E R C

**Principal Investigator**   Susan Jo Russell

**Co-Principal Investigator**   Cornelia C. Tierney

**Director of Research and Evaluation**   Jan Mokros

**Director of K–2 Curriculum**   Karen Economopoulos

**Curriculum Development**
Karen Economopoulos
Marlene Kliman
Jan Mokros
Megan Murray
Susan Jo Russell
Tracey Wright

**Evaluation and Assessment**
Mary Berle-Carman
Jan Mokros
Andee Rubin

**Teacher Support**
Irene Baker
Megan Murray
Judy Storeygard
Tracey Wright

**Technology Development**
Michael T. Battista
Douglas H. Clements
Julie Sarama

**Video Production**
David A. Smith
Judy Storeygard

**Administration and Production**
Irene Baker
Amy Catlin

**Cooperating Classrooms
for This Unit**
Malia Scott
Brookline Public Schools
Brookline, MA

Maryellen Bertrand
Boston Public Schools
Boston, MA

Elizabeth A. Pedrini
Arlington Public Schools
Arlington, MA

**Consultants and Advisors**
Deborah Lowenberg Ball
Michael T. Battista
Marilyn Burns
Douglas H. Clements
Ann Grady

# CONTENTS

## WHERE TO START

The first-time user of *Building Number Sense* should read the following:

When you next teach this same unit, you can begin to read more of the background. Each time you present the unit, you will learn more about how your students understand the mathematical ideas.

*Investigations in Number, Data, and Space®* is a K–5 mathematics curriculum with four major goals:

- to offer students meaningful mathematical problems
- to emphasize depth in mathematical thinking rather than superficial exposure to a series of fragmented topics
- to communicate mathematics content and pedagogy to teachers
- to substantially expand the pool of mathematically literate students

The *Investigations* curriculum embodies a new approach based on years of research about how children learn mathematics. Each grade level consists of a set of separate units, each offering 2–8 weeks of work. These units of study are presented through investigations that involve students in the exploration of major mathematical ideas.

Approaching the mathematics content through investigations helps students develop flexibility and confidence in approaching problems, fluency in using mathematical skills and tools to solve problems, and proficiency in evaluating their solutions. Students also build a repertoire of ways to communicate about their mathematical thinking, while their enjoyment and appreciation of mathematics grows.

The investigations are carefully designed to invite all students into mathematics—girls and boys, members of diverse cultural, ethnic, and language groups, and students with different strengths and interests. Problem contexts often call on students to share experiences from their family, culture, or community. The curriculum eliminates barriers—such as work in isolation from peers, or emphasis on speed and memorization—that exclude some students from participating successfully in mathematics. The following aspects of the curriculum ensure that all students are included in significant mathematics learning:

- Students spend time exploring problems in depth.
- They find more than one solution to many of the problems they work on.
- They invent their own strategies and approaches, rather than relying on memorized procedures.
- They choose from a variety of concrete materials and appropriate technology, including calculators, as a natural part of their everyday mathematical work.
- They express their mathematical thinking through drawing, writing, and talking.
- They work in a variety of groupings—as a whole class, individually, in pairs, and in small groups.
- They move around the classroom as they explore the mathematics in their environment and talk with their peers.

While reading and other language activities are typically given a great deal of time and emphasis in elementary classrooms, mathematics often does not get the time it needs. If students are to experience mathematics in depth, they must have enough time to become engaged in real mathematical problems. We believe that a minimum of five hours of mathematics classroom time a week—about an hour a day—is critical at the elementary level. The plan and pacing of the *Investigations* curriculum is based on that belief.

We explain more about the pedagogy and principles that underlie these investigations in Teacher Notes throughout the units. For correlations of the curriculum to the NCTM Standards and further help in using this research-based program for teaching mathematics, see the following books:

- *Implementing the* Investigations in Number, Data, and Space® *Curriculum*
- *Beyond Arithmetic: Changing Mathematics in the Elementary Classroom* by Jan Mokros, Susan Jo Russell, and Karen Economopoulos

This book is one of the curriculum units for *Investigations in Number, Data, and Space.* In addition to providing part of a complete mathematics curriculum for your students, this unit offers information to support your own professional development. You, the teacher, are the person who will make this curriculum come alive in the classroom; the book for each unit is your main support system.

Although the curriculum does not include student textbooks, reproducible sheets for student work are provided in the unit and are also available as Student Activity Booklets. Students work actively with objects and experiences in their own environment and with a variety of manipulative materials and technology, rather than with a book of instruction and problems. We strongly recommend use of the overhead projector as a way to present problems, to focus group discussion, and to help students share ideas and strategies.

Ultimately, every teacher will use these investigations in ways that make sense for his or her particular style, the particular group of students,

and the constraints and supports of a particular school environment. Each unit offers information and guidance for a wide variety of situations, drawn from our collaborations with many teachers and students over many years. Our goal in this book is to help you, a professional educator, implement this curriculum in a way that will give all your students access to mathematical power.

## Investigation Format

The opening two pages of each investigation help you get ready for the work that follows.

**What Happens** This gives a synopsis of each session or block of sessions.

**Mathematical Emphasis** This lists the most important ideas and processes students will encounter in this investigation.

**What to Plan Ahead of Time** These lists alert you to materials to gather, sheets to duplicate, transparencies to make, and anything else you need to do before starting.

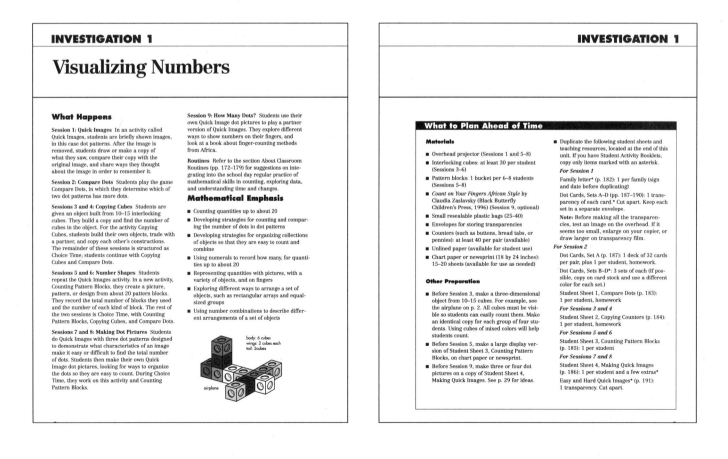

**INVESTIGATION 1**

# Visualizing Numbers

### What Happens

**Session 1: Quick Images** In an activity called Quick Images, students are briefly shown images, in this case dot patterns. After the image is removed, students draw or make a copy of what they saw, compare their copy with the original image, and share ways they thought about the image in order to remember it.

**Session 2: Compare Dots** Students play the game Compare Dots, in which they determine which of two dot patterns has more dots.

**Sessions 3 and 4: Copying Cubes** Students are given an object built from 10–15 interlocking cubes. They build a copy and find the number of cubes in the object. For the activity Copying Cubes, students build their own objects, trade with a partner, and copy each other's constructions. The remainder of these sessions is structured as Choice Time; students continue with Copying Cubes and Compare Dots.

**Sessions 5 and 6: Number Shapes** Students repeat the Quick Images activity. In a new activity, Counting Pattern Blocks, they create a picture, pattern, or design from about 20 pattern blocks. They record the total number of blocks they used and the number of each kind of block. The rest of the two sessions is Choice Time, with Counting Pattern Blocks, Copying Cubes, and Compare Dots.

**Sessions 7 and 8: Making Dot Pictures** Students do Quick Images with three dot patterns designed to demonstrate what characteristics of an image make it easy or difficult to find the total number of dots. Students then make their own Quick Image dot pictures, looking for ways to organize the dots so they are easy to count. During Choice Time, they work on this activity and Counting Pattern Blocks.

**Session 9: How Many Dots?** Students use their own Quick Image dot pictures to play a partner version of Quick Images. They explore different ways to show numbers on their fingers, and look at a book about finger-counting methods from Africa.

**Routines** Refer to the section About Classroom Routines (pp. 172–179) for suggestions on integrating into the school day regular practice of mathematical skills in counting, exploring data, and understanding time and changes.

### Mathematical Emphasis

- Counting quantities up to about 20
- Developing strategies for counting and comparing the number of dots in dot patterns
- Developing strategies for organizing collections of objects so that they are easy to count and combine
- Using numerals to record how many, for quantities up to about 20
- Representing quantities with pictures, with a variety of objects, and on fingers
- Exploring different ways to arrange a set of objects, such as rectangular arrays and equal-sized groups
- Using number combinations to describe different arrangements of a set of objects

body: 6 cubes
wings: 2 cubes each
tail: 2cubes

airplane

**INVESTIGATION 1**

### What to Plan Ahead of Time

#### Materials

- Overhead projector (Sessions 1 and 5–8)
- Interlocking cubes: at least 30 per student (Sessions 3–6)
- Pattern blocks: 1 bucket per 6–8 students (Sessions 5–8)
- *Count on Your Fingers African Style* by Claudia Zaslavsky (Black Butterfly Children's Press, 1996) (Session 9, optional)
- Small resealable plastic bags (25–40)
- Envelopes for storing transparencies
- Counters (such as buttons, bread tabs, or pennies): at least 40 per pair (available)
- Unlined paper (available for student use)
- Chart paper or newsprint (18 by 24 inches): 15–20 sheets (available for use as needed)

#### Other Preparation

- Before Session 3, make a three-dimensional object from 10–15 cubes. For example, see the airplane on p. 2. All cubes must be visible so students can easily count them. Make an identical copy for each group of four students. Using cubes of mixed colors will help students count.
- Before Session 5, make a large display version of Student Sheet 3, Counting Pattern Blocks, on chart paper or newsprint.
- Before Session 9, make three or four dot pictures on a copy of Student Sheet 4, Making Quick Images. See p. 29 for ideas.

- Duplicate the following student sheets and teaching resources, located at the end of this unit. If you have Student Activity Booklets, copy only items marked with an asterisk.

  *For Session 1*

  Family letter* (p. 182): 1 per family (sign and date before duplicating)

  Dot Cards, Sets A–D (pp. 187–190): 1 transparency of each card.* Cut apart. Keep each set in a separate envelope.

  **Note:** Before making all the transparencies, test an image on the overhead. If it seems too small, enlarge on your copier, or draw larger on transparency film.

  *For Session 2*

  Dot Cards, Set A (p. 187): 1 deck of 32 cards per pair, plus 1 per student, homework.

  Dot Cards, Sets B–D*: 3 sets of each (If possible, copy on card stock and use a different color for each set.)

  Student Sheet 1, Compare Dots (p. 183): 1 per student, homework.

  *For Sessions 3 and 4*

  Student Sheet 2, Copying Counters (p. 184): 1 per student, homework.

  *For Sessions 5 and 6*

  Student Sheet 3, Counting Pattern Blocks (p. 185): 1 per student

  *For Sessions 7 and 8*

  Student Sheet 4, Making Quick Images (p. 186): 1 per student and a few extras*

  Easy and Hard Quick Images* (p. 191): 1 transparency. Cut apart.

**Sessions** Within an investigation, the activities are organized by class session, a session being at least a one-hour math class. Sessions are numbered consecutively through an investigation. Often several sessions are grouped together, presenting a block of activities with a single major focus.

When you find a block of sessions presented together—for example, Sessions 1, 2, and 3— read through the entire block first to understand the overall flow and sequence of the activities. Make some preliminary decisions about how you will divide the activities into three sessions for your class, based on what you know about your students. You may need to modify your initial plans as you progress through the activities, and you may want to make notes in the margins of the pages as reminders for the next time you use the unit.

Be sure to read the Session Follow-Up section at the end of the session block to see what homework assignments and extensions are suggested as you make your initial plans.

While you may be used to a curriculum that tells you exactly what each class session should cover, we have found that the teacher is in a better position to make these decisions. Each unit is flexible and may be handled somewhat differently by every teacher. While we provide guidance for how many sessions a particular group of activities is likely to need, we want you to be active in determining an appropriate pace and the best transition points for your class. It is not unusual for a teacher to spend more or less time than is proposed for the activities.

**Activities** The activities include pair and small-group work, individual tasks, and whole-class discussions. In any case, students are seated together, talking and sharing ideas during all work times. Students most often work cooperatively, although each student may record work individually.

**Choice Time** In most units, some sessions are structured with activity choices. In these cases, students may work simultaneously on different activities focused on the same mathematical ideas. Students choose which activities they want to do, and they cycle through them. You will need to decide how to set up and introduce these activities and how to let students make their choices. Some

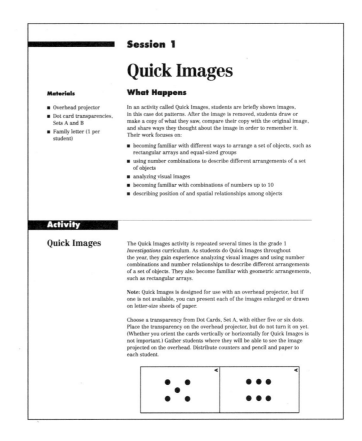

**Session 1**

# Quick Images

**Materials**

- Overhead projector
- Dot card transparencies, Sets A and B
- Family letter (1 per student)

**What Happens**

In an activity called Quick Images, students are briefly shown images, in this case dot patterns. After the image is removed, students draw or make a copy of what they saw, compare their copy with the original image, and share ways they thought about the image in order to remember it. Their work focuses on:

- becoming familiar with different ways to arrange a set of objects, such as rectangular arrays and equal-sized groups
- using number combinations to describe different arrangements of a set of objects
- analyzing visual images
- becoming familiar with combinations of numbers up to 10
- describing position of and spatial relationships among objects

**Activity**

**Quick Images**

The Quick Images activity is repeated several times in the grade 1 *Investigations* curriculum. As students do Quick Images throughout the year, they gain experience analyzing visual images and using number combinations and number relationships to describe different arrangements of a set of objects. They also become familiar with geometric arrangements, such as rectangular arrays.

**Note:** Quick Images is designed for use with an overhead projector, but if one is not available, you can present each of the images enlarged or drawn on letter-size sheets of paper.

Choose a transparency from Dot Cards, Set A, with either five or six dots. Place the transparency on the overhead projector, but do not turn it on yet. (Whether you orient the cards vertically or horizontally for Quick Images is not important.) Gather students where they will be able to see the image projected on the overhead. Distribute counters and pencil and paper to each student.

teachers set up choices as stations around the room, while others post the list of available choices and allow students to collect their own materials and choose their own work space. You may need to experiment with a few different structures before finding a setup that works best for you.

**Extensions** These follow-up activities are opportunities for some or all students to explore a topic in greater depth or in a different context. They are not designed for "fast" students; mathematics is a multi-faceted discipline, and different students will want to go further in different investigations. Look for and encourage the sparks of interest and enthusiasm you see in your students, and use the extensions to help them pursue these interests.

**Excursions** Some of the *Investigations* units include excursions—blocks of activities that could be omitted without harming the integrity of the unit. This is one way of dealing with the great depth and variety of elementary mathematics— much more than a class has time to explore in any one year. Excursions give you the flexibility to make different choices from year to year, doing the excursion in one unit this time, and next year trying another excursion.

**Tips for the Linguistically Diverse Classroom**  At strategic points in each unit, you will find concrete suggestions for simple modifications of the teaching strategies to encourage the participation of all students. Many of these tips offer alternative ways to elicit critical thinking from students at varying levels of English proficiency, as well as from other students who find it difficult to verbalize their thinking.

The tips are supported by suggestions for specific vocabulary work to help ensure that all students can participate fully in the investigations. The Preview for the Linguistically Diverse Classroom (p. I-22) lists important words that are assumed as part of the working vocabulary of the unit. Second-language learners will need to become familiar with these words in order to understand the problems and activities they will be doing. These terms can be incorporated into students' second-language work before or during the unit. Activities that can be used to present the words are found in the appendix, Vocabulary Support for Second-Language Learners (p. 180). In addition, ideas for making connections to students' language and cultures, included on the Preview page, help the class explore the unit's concepts from a multi-cultural perspective.

**Classroom Routines**  Activities in counting, exploring data, and understanding time and changes are suggested for routines in the grade 1 *Investigations* curriculum. Routines offer ongoing work with this important content as a regular part of the school day. Some routines provide more practice with content presented in the curriculum; others extend the curriculum; still others explore new content areas.

Plan to incorporate a few of the routine activities into a standard part of your daily schedule, such as morning meeting. When opportunities arise, you can also include routines as part of your work in other subject areas (for example, keeping a weather chart for science). Most routines are short and can be done whenever you have a spare 10–15 minutes, such as before lunch or recess or at the end of the day.

You will need to decide how often to present routines, what variations are appropriate for your class, and at what points in the day or week you will include them. A reminder about classroom routines is included on the first page of each investigation. Whatever routines you choose, your students will gain the most from these routines if they work with them regularly.

## Materials

A complete list of the materials needed for teaching this unit is found on p. I-17. Some of these materials are available in kits for the *Investigations* curriculum. Individual items can also be purchased from school supply dealers.

**Classroom Materials**  In an active mathematics classroom, certain basic materials should be available at all times: interlocking cubes, pencils, unlined paper, graph paper, calculators, and things to count with. Some activities in this curriculum require scissors and glue sticks or tape. Stick-on notes and large paper are also useful materials throughout.

So that students can independently get what they need at any time, they should know where these materials are kept, how they are stored, and how they are to be returned to the storage area. Many teachers have found that stopping 5 minutes before the end of each session so that students can finish their work and clean up is helpful in maintaining classroom materials. You'll find that establishing such routines at the beginning of the year is well worth the time and effort.

**Technology** Calculators are introduced to students in the first unit of the grade 1 sequence, *Mathematical Thinking at Grade 1.* By freely exploring and experimenting, students become familiar with this important mathematical tool.

Computer activities at grade 1 use a software program, called Shapes, that was developed especially for the *Investigations* curriculum. This program is introduced in the geometry unit, *Quilt Squares and Block Towns.* Using *Shapes,* students explore two-dimensional geometry while making pictures and designs with pattern block shapes and tangram pieces.

Although the software is linked to activities only in the geometry unit, we recommend that students use it throughout the year. Thus, you may want to introduce it when you introduce pattern blocks in *Mathematical Thinking at Grade 1.* How you use the computer activities depends on the number of computers you have available. Suggestions are offered in the geometry unit for how to organize different types of computer environments.

**Children's Literature** Each unit offers a list of suggested children's literature (p. I-17) that can be used to support the mathematical ideas in the unit. Sometimes an activity is based on a specific children's book, with suggestions for substitutions where practical. While such activities can be adapted and taught without the book, the literature offers a rich introduction and should be used whenever possible.

**Student Sheets and Teaching Resources** Student recording sheets and other teaching tools needed for both class and homework are provided as reproducible blackline masters at the end of each

---

**STORY PROBLEMS, SET A**

Copy one set per student. Cut apart and place copies of each problem in a separate envelope.

| | |
|---|---|
| 1. Rosa had 6 toy cars. Her mom gave her 6 more. How many toy cars does Rosa have now?<br><br>A | 2. Steve has 5 marbles in his pocket. He has 9 marbles in a bag. How many marbles does he have?<br><br>A |
| 3. A squirrel ate 8 nuts. Then she ate 7 more nuts. How many nuts did she eat?<br><br>A | 4. Rosa had 11 library books. She took 4 of them back. Now how many books does she have??<br><br>A |
| 5. Steve and Rosa had 12 apples. They ate 6 of them. How many apples are left?<br><br>A | 6. Rosa had 15 pennies. She spent 6 pennies. How many pennies did she have then?<br><br>A |

---

unit. They are also available as Student Activity Booklets. These booklets contain all the sheets each student will need for individual work, freeing you from extensive copying (although you may need or want to copy the occasional teaching resource on transparency film or card stock, or make extra copies of a student sheet).

We think it's important that students find their own ways of organizing and recording their work. They need to learn how to explain their thinking with both drawings and written words, and how to organize their results so someone else can understand them. For this reason, we deliberately do not provide student sheets for every activity. Regardless of the form in which students do their work, we recommend that they keep a mathematics notebook or folder so that their work is always available for reference.

**Homework** In *Investigations,* homework is an extension of classroom work. Sometimes it offers review and practice of work done in class, sometimes preparation for upcoming activities, and sometimes numerical practice that revisits work in earlier units. Homework plays a role both in supporting students' learning and in helping inform families about the ways in which students in this curriculum work with mathematical ideas.

Depending on your school's homework policies and your own judgment, you may want to assign more homework than is suggested in the units. For this purpose you might use the practice pages, included as blackline masters at the end of this unit, to give students additional work with numbers.

For some homework assignments, you will want to adapt the activity to meet the needs of a variety of students in your class: those with special needs, those ready for more challenge, and second-language learners. You might change the numbers in a problem, make the activity more or less complex, or go through a sample activity with those who need extra help. You can modify any student sheet for either homework or class use. In particular, making numbers in a problem smaller or larger can make the same basic activity appropriate for a wider range of students.

Another issue to consider is how to handle the homework that students bring back to class—how to recognize the work they have done at home without spending too much time on it. Some teachers hold a short group discussion of different approaches to the assignment; others ask students to share and discuss their work with a neighbor, or post the homework around the room and give students time to tour it briefly. If you want to keep track of homework students bring in, be sure it ends up in a designated place.

*Investigations* **at Home** It is a good idea to make your policy on homework explicit to both students and their families when you begin teaching with *Investigations.* How frequently will you be assigning homework? When do you expect homework to be completed and brought back to school? What are your goals in assigning homework? How independent should families expect their children to be? What should the parent or guardian's role be? The more explicit you can be about your expectations, the better the homework experience will be for everyone.

*Investigations* at Home (a booklet available separately for each unit, to send home with students) gives you a way to communicate with families about the work students are doing in class. This booklet includes a brief description of every session, a list of the mathematics content emphasized in each investigation, and a discussion of each homework assignment to help families more effectively support their children. Whether or not you are using the *Investigations* at Home booklets, we expect you to make your own choices about homework assignments. Feel free to omit any and to add extra ones you think are appropriate.

**Family Letter** A letter that you can send home to students' families is included with the blackline masters for each unit. Families need to be informed about the mathematics work in your classroom; they should be encouraged to participate in and support their children's work. A reminder to send home the letter for each unit appears in one of the early investigations. These letters are also available separately in Spanish, Vietnamese, Cantonese, Hmong, and Cambodian.

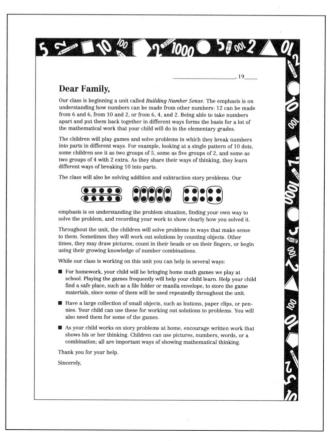

## Help for You, the Teacher

Because we believe strongly that a new curriculum must help teachers think in new ways about mathematics and about their students' mathematical thinking processes, we have included a great deal of material to help you learn more about both.

**About the Mathematics in This Unit** This introductory section (p. I-18) summarizes the critical information about the mathematics you will be teaching. It describes the unit's central mathematical ideas and how students will encounter them through the unit's activities.

**Teacher Notes** These reference notes provide practical information about the mathematics you are teaching and about our experience with how students learn. Many of the notes were written in response to actual questions from teachers, or to discuss important things we saw happening in the field-test classrooms. Some teachers like to read them all before starting the unit, then review them as they come up in particular investigations.

**Dialogue Boxes** Sample dialogues demonstrate how students typically express their mathematical ideas, what issues and confusions arise in their thinking, and how some teachers have guided class discussions.

These dialogues are based on the extensive classroom testing of this curriculum; many are word-for-word transcriptions of recorded class discussions. They are not always easy reading; sometimes it may take some effort to unravel what the students are trying to say. But this is the value of these dialogues; they offer good clues to how your students may develop and express their approaches and strategies, helping you prepare for your own class discussions.

**Where to Start** You may not have time to read everything the first time you use this unit. As a first-time user, you will likely focus on understanding the activities and working them out with your students. Read completely through each investigation before starting to present it. Also read those sections listed in the Contents (p. vi) under the heading Where to Start.

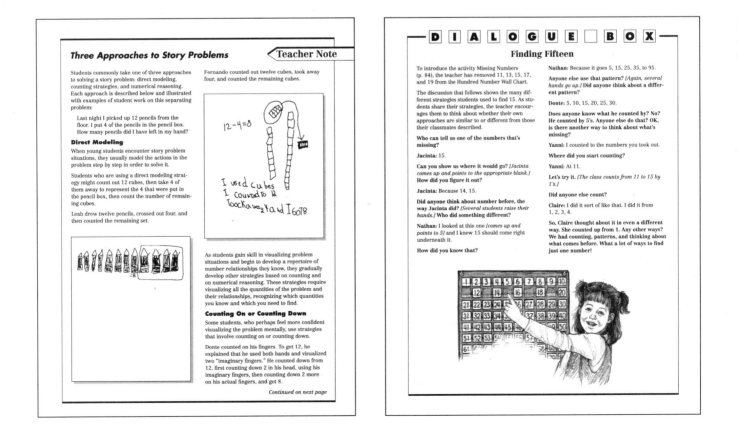

The *Investigations* curriculum incorporates the use of two forms of technology in the classroom: calculators and computers. Calculators are assumed to be standard classroom materials, available for student use in any unit. Computers are explicitly linked to one or more units at each grade level; they are used with the unit on 2-D geometry unit at each grade, as well as with some of the units on measuring, data, and changes.

## Using Calculators

In this curriculum, calculators are considered tools for doing mathematics, similar to pattern blocks or interlocking cubes. Just as with other tools, students must learn both *how* to use calculators correctly and *when* they are appropriate to use. This knowledge is crucial for daily life, as calculators are now a standard way of handling numerical operations, both at work and at home.

Using a calculator correctly is not a simple task; it depends on a good knowledge of the four operations and of the number system, so that students can select suitable calculations and also determine what a reasonable result would be. These skills are the basis of any work with numbers, whether or not a calculator is involved.

Unfortunately, calculators are often seen as tools to check computations with, as if other methods are somehow more fallible. Students need to understand that any computational method can be used to check any other; it's just as easy to make a mistake on the calculator as it is to make a mistake on paper or with mental arithmetic. Throughout this curriculum, we encourage students to solve computation problems in more than one way in order to double-check their accuracy. We present mental arithmetic, paper-and-pencil computation, and calculators as three possible approaches.

In this curriculum we also recognize that, despite their importance, calculators are not always appropriate in mathematics instruction. Like any tools, calculators are useful for some tasks, but not for others. You will need to make decisions about when to allow students access to calculators and when to ask that they solve problems without them, so that they can concentrate on other tools and skills. At times when calculators are or are not appropriate for a particular activity, we make specific recommendations. Help your students develop their own sense of which problems they can tackle with their own reasoning and which ones might be better solved with a combination of their own reasoning and the calculator.

Managing calculators in your classroom so that they are a tool, and not a distraction, requires some planning. When calculators are first introduced, students often want to use them for everything, even problems that can be solved quite simply by other methods. However, once the novelty wears off, students are just as interested in developing their own strategies, especially when these strategies are emphasized and valued in the classroom. Over time, students will come to recognize the ease and value of solving problems mentally, with paper and pencil, or with manipulatives, while also understanding the power of the calculator to facilitate work with larger numbers.

Experience shows that if calculators are available only occasionally, students become excited and distracted when they are permitted to use them. They focus on the tool rather than on the mathematics. In order to learn when calculators are appropriate and when they are not, students must have easy access to them and use them routinely in their work.

If you have a calculator for each student, and if you think your students can accept the responsibility, you might allow them to keep their calculators with the rest of their individual materials, at least for the first few weeks of school. Alternatively, you might store them in boxes on a shelf, number each calculator, and assign a corresponding number to each student. This system can give students a sense of ownership while also helping you keep track of the calculators.

## Using Computers

Students can use computers to approach and visualize mathematical situations in new ways. The computer allows students to construct and manipulate geometric shapes, see objects move according to rules they specify, and turn, flip, and repeat a pattern.

This curriculum calls for computers in units where they are a particularly effective tool for learning mathematics content. One unit on 2-D geometry at each of the grades 3–5 includes a core of activities that rely on access to computers, either in the classroom or in a lab. Other units on geometry, measurement, data, and changes include computer activities, but can be taught without them. In these units, however, students' experience is greatly enhanced by computer use.

The following list outlines the recommended use of computers in this curriculum:

**Grade 1**
Unit: *Survey Questions and Secret Rules* (Collecting and Sorting Data)
Software: Tabletop, Jr.
Source: Broderbund

Unit: *Making Triangles and Building Cities* (2-D and 3-D Geometry)
Software: *Shapes*
Source: provided with the unit

**Grade 2**
Unit: *Mathematical Thinking at Grade 2* (Introduction)
Software: *Shapes*
Source: provided with the unit

Unit: *Shapes, Halves, and Symmetry* (Geometry and Fractions)
Software: Shapes
Source: provided with the unit

Unit: *How Long? How Far?* (Measuring)
Software: *Geo-Logo*
Source: provided with the unit

**Grade 3**
Unit: *Flips, Turns, and Area* (2-D Geometry)
Software: *Tumbling Tetrominoes*
Source: provided with the unit

Unit: *Turtle Paths* (2-D Geometry)
Software: *Geo-Logo*
Source: provided with the unit

**Grade 4**
Unit: *Sunken Ships and Grid Patterns* (2-D Geometry)
Software: *Geo-Logo*
Source: provided with the unit

**Grade 5**
Unit: *Picturing Polygons* (2-D Geometry)
Software: *Geo-Logo*
Source: provided with the unit

Unit: *Patterns of Change* (Tables and Graphs)
Software: *Trips*
Source: provided with the unit

Unit: *Data: Kids, Cats, and Ads* (Statistics)
Software: Tabletop, Sr.
Source: Broderbund

The software provided with the *Investigations* units uses the power of the computer to help students explore mathematical ideas and relationships. With the *Shapes* (grades 1–2) and *Tumbling Tetrominoes* (grade 3) software, students explore symmetry, pattern, rotation and reflection, area, and characteristics of 2-D shapes. With the *Geo-Logo* software (grades 3–5), students investigate rotations and reflections, coordinate geometry, the properties of 2-D shapes, and angles. The *Trips* software (grade 5) is a mathematical exploration of motion in which students run experiments and interpret data presented in graphs and tables.

We suggest that students work in pairs on the computer; this not only maximizes computer resources but also encourages students to consult, monitor, and teach one another. Generally, more than two students at one computer find it difficult to share. Managing access to computers is an issue for every classroom. The curriculum gives you explicit support for setting up a system. The units are structured on the assumption that you have enough computers for half your students to work on the machines in pairs at one time. If you do not have access to that many computers, suggestions are made for structuring class time to use the unit with five to eight computers, or even with fewer than five.

Assessment plays a critical role in teaching and learning, and it is an integral part of the *Investigations* curriculum. For a teacher using these units, assessment is an ongoing process. You observe students' discussions and explanations of their strategies on a daily basis and examine their work as it evolves. While students are busy recording and representing their work, working on projects, sharing with partners, and playing mathematical games, you have many opportunities to observe their mathematical thinking. What you learn through observation guides your decisions about how to proceed. In any of the units, you will repeatedly consider questions like these:

- Do students come up with their own strategies for solving problems, or do they expect others to tell them what to do? What do their strategies reveal about their mathematical understanding?

- Do students understand that there are different strategies for solving problems? Do they articulate their strategies and try to understand other students' strategies?

- How effectively do students use materials as tools to help with their mathematical work?

- Do students have effective ideas for keeping track of and recording their work? Does keeping track of and recording their work seem difficult for them?

You will need to develop a comfortable and efficient system for recording and keeping track of your observations. Some teachers keep a clipboard handy and jot notes on a class list or on adhesive labels that are later transferred to student files. Others keep loose-leaf notebooks with a page for each student and make weekly notes about what they have observed in class.

## Assessment Tools in the Unit

With the activities in each unit, you will find questions that help focus your observations. You will also find two built-in assessment tools: Teacher Checkpoints and embedded Assessment activities.

**Teacher Checkpoints**  The designated Teacher Checkpoints in each unit offer a time to "check in" with individual students, watch them at work, and ask questions that illuminate how they are thinking.

At first it may be hard to know what to look for, hard to know what kinds of questions to ask. Students may be reluctant to talk; they may not be accustomed to having the teacher ask them about their work, or they may not know how to explain their thinking. Two important ingredients of this process are asking students open-ended questions about their work and showing genuine interest in how they are approaching the task. When students see that you are interested in their thinking and are counting on them to come up with their own ways of solving problems, they may surprise you with the depth of their understanding.

Teacher Checkpoints also give you the chance to pause in the teaching sequence and reflect on how your class is doing overall. Think about whether you need to adjust your pacing: Are most students fluent with strategies for solving a particular kind of problem? Are they just starting to formulate good strategies? Or are they still struggling with how to start? Depending on what you see as the students work, you may want to spend more time on similar problems, change some of the problems to use smaller numbers, move quickly to more challenging material, modify subsequent activities for some students, work on particular ideas with a small group, or pair students who have good strategies with those who are having more difficulty.

**Embedded Assessment Activities**  Assessment activities embedded in each unit will help you examine specific pieces of student work, figure out what it means, and provide feedback. From the students' point of view, these assessment activities are no different from any others. Each is a learning experience in and of itself, as well as an opportunity for you to gather evidence about students' mathematical understanding.

The embedded assessment activities sometimes involve writing and reflecting; at other times, a discussion or brief interaction between student and teacher; and in still other instances, the creation and explanation of a product. In most cases, the assessments require that students show what they did, write or talk about it, or do both. Having to explain how they worked through a problem helps students be more focused and clear in their mathematical thinking. It also helps them realize that doing mathematics is a process that may involve tentative starts, revising one's approach, taking different paths, and working through ideas.

Teachers often find the hardest part of assessment to be interpreting their students' work. We provide guidelines to help with that interpretation. If you have used a process approach to teaching writing, the assessment in *Investigations* will seem familiar. For many of the assessment activities, a Teacher Note provides examples of student work and a commentary on what it indicates about student thinking.

## Documentation of Student Growth

To form an overall picture of mathematical progress, it is important to document each student's work in journals, notebooks, or portfolios. The choice is largely a matter of personal preference; some teachers have students keep a notebook or folder for each unit, while others prefer one mathematics notebook, or a portfolio of selected work for the entire year. The final activity in each *Investigations* unit, called Choosing Student Work to Save, helps you and the students select representative samples for a record of their work.

This kind of regular documentation helps you synthesize information about each student as a mathematical learner. From different pieces of evidence, you can put together the big picture. This synthesis will be invaluable in thinking about where to go next with a particular child, deciding where more work is needed, or explaining to parents (or other teachers) how a child is doing.

If you use portfolios, you need to collect a good balance of work, yet avoid being swamped with an overwhelming amount of paper. Following are some tips for effective portfolios:

- Collect a representative sample of work, including some pieces that students themselves select for inclusion in the portfolio. There should be just a few pieces for each unit, showing different kinds of work—some assignments that involve writing, as well as some that do not.

- If students do not date their work, do so yourself so that you can reconstruct the order in which pieces were done.

- Include your reflections on the work. When you are looking back over the whole year, such comments are reminders of what seemed especially interesting about a particular piece; they can also be helpful to other teachers and to parents. Older students should be encouraged to write their own reflections about their work.

## Assessment Overview

There are two places to turn for a preview of the assessment opportunities in each *Investigations* unit. The Assessment Resources column in the unit Overview Chart (pp. I-13–I-16) identifies the Teacher Checkpoints and Assessment activities embedded in each investigation, guidelines for observing the students that appear within classroom activities, and any Teacher Notes and Dialogue Boxes that explain what to look for and what types of student responses you might expect to see in your classroom. Additionally, the section About the Assessment in This Unit (p. I-20) gives you a detailed list of questions for each investigation, keyed to the mathematical emphases, to help you observe student growth.

Depending on your situation, you may want to provide additional assessment opportunities. Most of the investigations lend themselves to more frequent assessment, simply by having students do more writing and recording while they are working.

# Building Number Sense

**Content of This Unit**  Students discover ways that numbers can be made from other numbers; that is, 12 can be made from 6 and 6, from 10 and 2, or from 6, 4, and 2. Being able to take numbers apart and put them back together flexibly is the basis for developing good number sense and an understanding of the operations. In this unit, students learn about numbers in lots of ways: They play games and solve problems that involve breaking numbers into parts in different ways; they use numbers to describe dot patterns; they count sets of objects; they read, write, and sequence numbers up to 100; they find the total of two or more numbers; and they compare numbers to find which is larger. Students also use their growing number sense to develop strategies for solving addition and subtraction story problems, finding their own way to solve the problem and to record their thinking.

Throughout the unit, students are encouraged to solve problems in ways that make sense to them. Sometimes they may work out solutions by using real objects, such as pennies, buttons, or paper clips. Others times, they may draw pictures, count in their heads or on their fingers, or begin using what they know about number and number relationships.

**Connections with Other Units**  If you are doing the full-year *Investigations* curriculum, this is the second of six units. Your class will already be familiar with several of the activities from their work in *Mathematical Thinking at Grade 1,* including How Many of Each? problems, Clapping Patterns, and the game Double Compare. They will have had some experience with number combinations, combining numbers, and counting. They will continue working on these and other important number ideas in the unit, *Number Games and Story Problems.*

## *Investigations* Curriculum ▪ Suggested Grade 1 Sequence

*Mathematical Thinking at Grade 1* (Introduction)

▶ *Building Number Sense* (The Number System)

*Survey Questions and Secret Rules* (Collecting and Sorting Data)

*Quilt Squares and Block Towns* (2-D and 3-D Geometry)

*Number Games and Story Problems* (Addition and Subtraction)

*Bigger, Taller, Heavier, Smaller* (Measuring)

# Investigation 1 ▪ Visualizing Numbers

| Class Sessions | Activities | Pacing |
|---|---|---|
| Session 1 (p. 4)<br>QUICK IMAGES | Quick Images<br>Homework: Family Connection | minimum<br>1 hr |
| Session 2 (p. 8)<br>COMPARE DOTS | Teacher Checkpoint: Compare Dots<br>Homework: Compare Dots<br>Extension: Double Compare Dots | minimum<br>1 hr |
| Sessions 3 and 4 (p. 13)<br>COPYING CUBES | Copying Cubes<br>Introducing Choice Time<br>Homework: Copying Counters | minimum<br>2 hr |
| Sessions 5 and 6 (p. 18)<br>NUMBER SHAPES | Quick Images<br>Counting Pattern Blocks<br>Choice Time | minimum<br>2 hr |
| Sessions 7 and 8 (p. 24)<br>MAKING DOT PICTURES | Easy and Hard Quick Images<br>Choice Time<br>Homework: Double Compare Dots | minimum<br>2 hr |
| Session 9 (p. 29)<br>HOW MANY DOTS? | How Many Dots?<br>Numbers on Our Fingers | minimum<br>1 hr |
| **Classroom Routines** (see pp. 172–179) | | |

## Mathematical Emphasis

- Counting quantities up to about 20

- Developing strategies for counting and comparing the number of dots in dot patterns (in which dots are arranged in rows or in groups)

- Developing strategies for organizing collections of objects so that they are easy to count and combine

- Using numerals to record how many, for quantities up to about 20

## Assessment Resources

Teacher Checkpoint: Compare Dots (p. 8)

Observing the Students (pp. 9, 16, 21, 27)

Finding the Number of Dots (Teacher Note, p. 11)

Observing Students Counting (Teacher Note, p. 23)

Easy and Hard Quick Images (Dialogue Box, p. 28)

## Materials

Interlocking cubes

Pattern blocks

*Count on Your Fingers African Style* (opt.)

Counters

Overhead projector

Resealable plastic bags

Envelopes

Unlined paper

Chart paper

Student Sheets 1–4

Teaching resource sheets

# Investigation 2 ▪ Building Numbers in Different Ways

| Class Sessions | Activities | Pacing |
|---|---|---|
| Session 1 (p. 38)<br>TWELVE CATS AND DOGS | Teacher Checkpoint: Twelve Cats and Dogs<br>Sharing Solutions<br>Homework: Turtles and Frogs<br>Extension: Finding All the Solutions<br>Extension: Cats, Dogs, and Rabbits<br>Extension: Ways to Make 12 | minimum<br>1 hr |
| Session 2 (p. 48)<br>PROBLEMS ABOUT TEN | How Many of Each? Problems About Ten<br>Sharing Our Problems About Ten<br>Extension: Finding All the Solutions<br>Extension: Combining Solutions<br>Extension: Larger Numbers | minimum<br>1 hr |
| Session 3 (p. 53)<br>THE GAME OF DOUBLE COMPARE | Double Compare<br>Homework: Double Compare<br>Extension: Triple Compare | minimum<br>1 hr |
| Sessions 4 and 5 (p. 59)<br>BREAKING NUMBERS INTO<br>TWO PARTS | Introducing Two New Games<br>Choice Time<br>Sharing Combinations of Seven<br>Homework: On and Off<br>Homework: Counters in a Cup<br>Extension: On and Off: Finding All the Ways | minimum<br>2 hr |
| Sessions 6, 7, and 8 (p. 68)<br>TOWERS OF 10 AND NUMBER<br>CHOICES | Introducing Towers of 10<br>Choice Time<br>Quick Images<br>Sharing Combinations of Ten<br>Homework: Math Games<br>Extension: All the Combinations of 10 | minimum<br>3 hr |
| Session 9 (p. 00)<br>THREE TOWERS | Assessment: Three Towers<br>Fifteen in All | minimum<br>1 hr |

**Classroom Routines** (see pp. 172–179)

| Mathematical Emphasis | Assessment Resources | Materials |
|---|---|---|
| ▪ Finding combinations of numbers up to about 15<br><br>▪ Finding the total of two quantities up to 10<br><br>▪ Finding the larger of two quantities up to about 20 | Teacher Checkpoint: Twelve Cats and Dogs (p. 38)<br><br>Observing the Students (pp. 39, 49, 54, 64, 71, 76)<br><br>Finding Relationships Among Solutions (Teacher Note, p. 45)<br><br>Double Compare: Strategies for Combining and Comparing (Teacher Note, p. 56)<br><br>Combinations of Seven (Dialogue Box, p. 67)<br><br>Assessment: Three Towers (p. 75)<br><br>Assessment: Three Towers (Teacher Note, p. 79) | Number Cards<br>Paper cups<br>Counters<br>Interlocking cubes<br>Dot cubes<br>Overhead projector<br>Unlined paper<br>Chart paper<br>Student Sheets 5–10<br>Teaching resource sheets |

# Investigation 3 ▪ Counting

| Class Sessions | Activities | Pacing |
|---|---|---|
| Sessions 1 and 2 (p. 82)<br>THE 100 CHART | Exploring the 100 Chart<br>Missing Numbers<br>Introducing Counting Strips<br>Choice Time<br>Homework: Numbers at Home | minimum<br>2 hr |
| Sessions 3 and 4 (p. 92)<br>WHICH HOLDS MORE? | Which Holds More?<br>Exploring Calculators<br>Choice Time<br>Homework: Math Games | minimum<br>2 hr |
| Sessions 5, 6, and 7 (p. 99)<br>MORE COUNTING AND COMPARING | Missing Numbers<br>Introducing Ten Turns<br>Choice Time<br>Discussing Our Counting Strips<br>Homework: Ten Turns | minimum<br>3 hr |
| Session 8 (p. 106)<br>CLAPPING PATTERNS | Clapping Patterns<br>Representing a Pattern | minimum<br>1 hr |
| Session 9 (Excursion)* (p. 110)<br>COUNTING STORIES | A Counting Story<br>A Class Counting Adventure<br>Extension: Are There 100? | minimum<br>1 hr |

**Classroom Routines** (see pp. 172–179)

*Excursions can be omitted without harming the integrity or continuity of the unit,
 but offer good mathematical work if you have time to include them.

## Mathematical Emphasis

- Reading, writing, and sequencing numbers to 100

- Counting quantities up to about 40

- Finding the total of two quantities, one that's just a few and another up to about 40

## Assessment Resources

Observing the Students (pp. 88, 96, 102, 109)

Exploring the 100 Chart (Dialogue Box, p. 90)

Finding Fifteen (Dialogue Box, p. 91)

What Went Wrong? (Dialogue Box, p. 105)

## Materials

Hundred Number Wall Chart

Colored plastic chart markers

Hundred Number Board with removable tiles

Interlocking cubes

Calculators

Number cubes

Plastic pennies

Pattern blocks

Counters

*Out for the Count: A Counting Adventure* (opt.)

Adding machine tape

Clear tape

Drawing paper

Crayons or markers

Unlined paper

Chart paper

Student Sheets 11–14

Teaching resource sheets

# Investigation 4 ■ Addition and Subtraction

| Class Sessions | Activities | Pacing |
|---|---|---|
| Session 1 (p. 116)<br>INTRODUCING<br>COMBINING SITUATIONS | Making Sense of Combining<br>Recording Combining Strategies<br>Sharing Strategies<br>Homework: Apples and Oranges | minimum<br>1 hr |
| Session 2 (p. 128)<br>INTRODUCING<br>SEPARATING SITUATIONS | Making Sense of Separating<br>Recording Separating Strategies<br>Sharing Separating Strategies<br>Homework: Eating Apples | minimum<br>1 hr |
| Sessions 3, 4, and 5 (p. 137)<br>FIVE-IN-A-ROW AND<br>STORY PROBLEMS | Five-in-a-Row<br>Choice Time<br>Sharing Story Problem Strategies<br>Teacher Checkpoint: Student Strategies<br>Homework: Finishing Story Probelms<br>Homework: Five-in-a-Row | minimum<br>3 hr |
| Session 6 (p. 148)<br>DOT ADDITION | Introducing Dot Addition<br>Playing Dot Addition<br>Homework: Dot Addition | minimum<br>1 hr |
| Sessions 7, 8, and 9 (p. 152)<br>COMBINING AND SEPARATING | Dot Addition Quick Images<br>Choice Time<br>Strategies for Finding Combinations<br>Sharing Story Problem Strategies<br>Homework: Story Problems<br>Homework: Dot Addition<br>Extension: What's a Story for This Problem? | minimum<br>3 hr |
| Session 10 (p. 160)<br>SOLVING STORY PROBLEMS | Assessment: Solving Story Problems<br>Clapping Patterns<br>Choosing Student Work to Save | minimum<br>1 hr |

**Classroom Routines** (see pp. 172–179)

## Mathematical Emphasis

- Visualizing combining and separating problem situations

- Developing strategies for solving combining and separating problems

- Recording strategies for solving combining and separating problems, using pictures, numbers, words, and equations

## Assessment Resources

Observing the Students (pp. 119, 140, 150, 155)

Recording Strategies for a Combining Problem (Dialogue Box, p. 126)

Three Approaches to Story Problems (Teacher Note, p. 133)

Teacher Checkpoint: Student Strategies (p. 143)

Assessment: Solving Story Problems (p. 160)

Assessment: Solving Story Problems (Teacher Note, p. 165)

## Materials

Number cubes

Overhead projector

Hundred Number Wall Chart

Paste or glue sticks

Counters

Unlined paper

Chart paper

Envelopes

Student Sheets 15–24

Teaching resource sheets

Following are the basic materials needed for the activities in this unit. Many of the items can be purchased from the publisher, either individually or in the Teacher Resource Package and the Student Materials Kit for grade 1. Detailed information is available on the *Investigations* order form. To obtain this form, call toll-free 1-800-872-1100 and ask for a Dale Seymour customer service representative.

Interlocking cubes (cubes that connect on all sides): a class supply with at least 30 per student

Pattern blocks: 1 bucket per 6–8 students

Counters, such as buttons, bread tabs, or pennies: at least 30 per student

Dot cubes (12–16 for the class)

Number cubes (12–16 for the class)

Play pennies (50–60 per pair, optional)

Calculators (at least 6–8 for the class)

Hundred Number Wall Chart (vinyl, with pockets), with numeral cards and chart markers (translucent colored plastic overlays)

Plastic Hundred Number Boards with removable number tiles (2 or 3 for the class)

*Out for the Count: A Counting Adventure* by Kathryn Cave (optional)

*Counting On Your Fingers African Style* by Claudia Zaslavsky (optional)

Adding machine tape (1 roll)

Paper cups (1 per student)

A dozen clean, nonbreakable containers that hold from 10 to 40 interlocking cubes (such as margarine tubs; paper or plastic cups and glasses; soup cans, coffee cans)

Unlined paper

Card stock

Chart paper or newsprint (18 by 24 inches)

Overhead projector and transparencies

Envelopes for storing overhead transparencies

Resealable plastic bags for storing cards

Paste or glue sticks

Tape

Crayons or markers

The following materials are provided at the end of this unit as blackline masters. A Student Activity Booklet containing all student sheets and teaching resources needed for individual work is available.

Family Letter (p. 182)

Student Sheets 1–24 (p. 183)

Teaching Resources:

   Dot Cards (p. 187)

   Easy and Hard Quick Images (p. 191)

   Number Cards (p. 198)

   100 Chart and Blank 100 Chart (p. 206)

   Cube Pattern Strips (p. 208)

   Dot Addition Cards (p. 220)

   Story Problems (p. 221)

   Game Record Sheet (p. 225)

Practice Pages (p. 227)

### Related Children's Literature

Anno, Mitsumasa. *Anno's Counting Book.* New York: Harper Collins, 1986.

Cave, Kathryn. *Out for the Count: A Counting Adventure.* New York: Simon and Schuster, 1992.

Fleming, Denise. *Count!* New York: Holt, 1992.

Kitchen, Bert. *Animal Numbers.* New York: Dial Books, 1987.

MacCarthy, Patricia. *The Ocean Parade: A Counting Book.* New York: Dial Books, 1990.

Sloat, Teri. *From One to One Hundred.* New York: A Puffin Unicorn Book, 1991.

Zaslavsky, Claudia. *Count on Your Fingers African Style.* New York: Black Butterfly Children's Press, 1996.

The focus of this unit is on developing and using good number sense. Good number sense includes (1) knowledge about how numbers are structured in our number system, (2) a large repertoire of known number relationships, and (3) a model of what it means to add and subtract. In this unit, young primary students are just beginning to build knowledge and experience in these three areas.

**The Number System**

Counting is a cornerstone of much of the number work that students do in first grade. Through frequent opportunities to count, students learn the oral number word sequence ("one, two, three, four, …") and the written number sequence (1, 2, 3, 4, …), and how these words and symbols represent quantities. At first, young children learn the oral counting sequence by rote and don't necessarily connect the numbers they say to a quantity being counted. In first grade, most students understand that one counting word is connected with one object, and that the last counting word names the quantity of objects being counted. However you may still have a few students who are working on this idea.

As students become comfortable with the early part of the oral counting sequence, they begin associating number words with the corresponding written numerals. They also become interested in counting higher and higher. In this unit, students extend and deepen their understanding of counting with numbers up to 100. As they count collections of objects, they become more familiar with the oral counting sequence, they gain experience keeping track of what has been counted and what is left to count, and they develop an understanding of the size of quantities.

As first graders read, say, and write numbers above 20, they begin to notice the patterns that reveal the structure of our number system. Work with the 100 chart gives students practice in reading, writing, and sequencing numbers; they also become familiar with patterns based on tens and ones. They notice that the thirties come after the twenties, the forties after the thirties, and so forth, and they come to see that the sequence within each decade has common features.

Through work with the 100 chart and other counting activities, students learn more about how numbers are related to quantities. Counting up or counting down by 1's is an important step in developing ideas about the number system: counting up 1 results in a number that is one more than the quantity you started with; counting down 1 gives you a quantity one less than the quantity you started with.

**Number Relationships**

As students become proficient at counting by 1's, they begin to develop a basic repertoire of number relationships. Consider a first grade How Many of Each? problem:

> There are 9 animals. Some are cats and some are dogs. How many could be cats and how many could be dogs?

Students may at first just find different combinations randomly. You will see them put out 7 blue cubes and 3 red cubes, then count them all, find out there are too many, and take a blue cube away. After recording this combination, they start all over again, putting out some red and blue cubes and adjusting until there are 9. However, later in the year, these same students might put out some blue cubes, find out there are 7, then figure out that they need 2 red cubes by counting on 8, 9. Still later, they "just know" that 7 plus 2 more makes 9. Either they remember the combination, or they are so confident and quick counting up 2 from 7 that they don't even realize they are doing it.

Gradually, students build up a set of number relationships that they automatically know. Ability to see numbers as composed of chunks of different sizes, rather than as a collection of ones, develops gradually over the early elementary years. In this unit, students work on visualizing how numbers can be broken up in different ways. For example, as they work with dot patterns, they see quantities arranged in rows or groups:

Some students see this picture as two groups of 5, some see it as five groups of 2, and some see it as two groups of 4 with 2 extra. As students share their ways of thinking about the picture, they become familiar with different combinations that make 10.

The dot pattern activities give students a visual model of number combinations. In other activities, students use stories, pictures, objects, and their fingers to find different ways of breaking numbers into parts. As they explore relationships among the number combinations they make, students gradually begin to rely less on counting by 1's. They may begin to "just know" some combinations of small numbers, such as some of the "doubles" (2 + 2, 3 + 3, 4 + 4) or combinations that make 10 (6 + 4, 2 + 8). Later in the year, some students may begin using one combination to figure out another ("I know 3 and 3 is 6, so 3 and 4 is 7"). Many will be able to quickly add on 1 or 2 to any number. Although most first graders will begin relying less on 1's in situations involving familiar numbers under about 10 or 12, they will continue counting by 1's in many situations, especially when larger numbers are involved.

## Addition and Subtraction

Throughout this unit, students use their growing understanding of number to solve problems that involve combining and separating quantities. An important emphasis is making sense of the problem situation. When young students are first introduced to story problems, they need to learn to visualize the actions in the problem, to recognize whether quantities are being combined or separated, and to determine whether the resulting quantity will be more or less than the initial one. Students can model the problem either mentally or with objects and actions.

Students are always encouraged to solve problems in ways that make sense to them. Many first graders will be most comfortable modeling the problem by acting it out in sequence. They combine 6 and 7, for example, by counting out a set of 6 objects, counting out a set of 7 objects, and then counting the total. These students may need to work with actual objects or pictures in order to have a sense of the goal of the problem and the quantities involved.

Other students might use strategies that involve counting. They might start at 7 and then count up 6, perhaps keeping track of the number they counted on their fingers. They do not need to count out each quantity separately and do not need to rely on pictures or objects to make sense of the problem.

Still other students will begin to use their knowledge of number combinations and flexible ways of thinking about number, rather than strategies involving counting. They might solve the problem by breaking 7 into 4 and 3, combining the 4 with the 6 because they "just know" that 6 and 4 is 10, and then counting up 3 more. All these approaches, singly or in combination, are appropriate for young students just beginning to solve addition and subtraction problems. It is essential that all students have the opportunity to construct approaches that are firmly grounded in their own developing number sense, that they can rely on to solve problems, and that they can explain clearly to others.

Recording work is a critical aspect of doing mathematics, both in order to keep track of procedures and to communicate methods and solutions to others. In this unit, students record their work using pictures, numbers, and words. As they solve problems, record their work, and share their strategies with you and with their classmates, they learn how to think through a problem carefully, model its actions, and double-check their solutions. They also learn to recognize and interpret standard notation for writing addition and subtraction equations.

**Mathematical Emphasis** At the beginning of each investigation, the Mathematical Emphasis section tells you what is most important for students to learn about during that investigation. Many of these understandings and processes are difficult and complex. Students gradually learn more and more about each idea over many years of schooling. Individual students will begin and end the unit with different levels of knowledge and skill, but all will begin to develop number sense and will become more comfortable talking about and showing their ways of solving mathematics problems.

Throughout the Investigations curriculum, there are many opportunities for ongoing daily assessment as you observe, listen to, and interact with students at work. In this unit you will find three Teacher Checkpoints:

Investigation 1, Session 2:
Compare Dots (p. 8)

Investigation 2, Session 1:
Twelve Cats and Dogs (p. 38)

Investigation 4, Sessions 3, 4, and 5:
Student Strategies (p. 143)

This unit also has two embedded assessment activities:

Investigation 2, Session 9:
Three Towers (p. 75)

Investigation 4, Session 10:
Solving Story Problems (p. 160)

In addition, you can use almost any activity in this unit to assess your students' needs and strengths. Following are questions to help you focus your observations in each investigation. You may want to keep track of your observations for each student to help you plan your curriculum and monitor students' growth. Suggestions for documenting student growth can be found in the section About Assessment (p. I-10).

## Investigation 1: Visualizing Numbers

■ Can students count a set of up to 20 objects accurately? Do they have ways to keep track of which objects they have counted? Do they know the number sequence? Do they have a sense of the size of quantities up to 20?

■ How do students count and compare dot patterns? Do they count one dot at a time? Do they start with the number in one group and then count up? Do they use knowledge of number combinations?

■ How do students count collections that include objects of two or more different shapes or colors? Do they count the objects in random order? Do they group the objects, for example by color, and then count them in groups? Do they use knowledge of number combinations to combine the number in each group?

■ Can they use numerals to record the number of objects in a group? Can they use zero (0) appropriately?

## Investigation 2: Building Numbers in Different Ways

■ How do students generate combinations of numbers equal to a given total? Do they take a random approach, combining numbers until they reach the total? Do they use strategies based on counting? strategies based on number combinations?

■ How do students combine two numbers? Do they use strategies that involve counting from 1? Do they count up from one of the numbers? Do they use knowledge of number combinations?

■ How do students compare two numbers? Do they count out objects for each number and then compare the sets? Do they know which numbers represent larger quantities?

## Investigation 3: Counting

■ What range of numbers are students comfortable reading, writing, and sequencing? Do students use patterns in the sequence of numbers to help them read, write, and sequence them?

■ Can students count a set of up to 40 objects accurately? Do they have a sense of the size of quantities up to 40?

■ What strategies do students use for combining a few objects with a set of up to about 40 objects? Do they recount the entire set from 1? Do they count on? Do they "just know" some sums? Can they quickly add on 1 or 2 to some numbers? to any number?

## Investigation 4: Addition and Subtraction

■ Are students comfortable with both combining and separating situations? Can they keep the situation in mind as they solve the problem? Can students choose a strategy that will work to solve the problem?

■ What strategies do students rely on for combining and separating? Do they count by 1's? Do they start with one quantity and count on or back? Do they use strategies that involve number combinations? Which students are taking numbers apart in ways that help them solve problems more easily?

■ Can students record their strategies for solving combining and separating problems in a way that makes sense, using some combination of pictures, words, and numbers?

In the *Investigations* curriculum, mathematical vocabulary is introduced naturally during the activities. We don't ask students to learn definitions of new terms; rather, they come to understand such words as *triangle, add, compare, data,* and *graph* by hearing them used frequently in discussion as they investigate new concepts. This approach is compatible with current theories of second-language acquisition, which emphasize the use of new vocabulary in meaningful contexts while students are actively involved with objects, pictures, and physical movement.

Listed below are some key words used in this unit that will not be new to most English speakers at this age level, but may be unfamiliar to students with limited English proficiency. You will want to spend additional time working on these words with your students who are learning English. If your students are working with a second-language teacher, you might enlist your colleague's aid in familiarizing students with these words, before and during this unit. In the classroom, look for opportunities for students to hear and use these words. Activities you can use to present the words are given in the appendix, Vocabulary Support for Second-Language Learners (p. 180).

**compare, more, less** As students build their understanding of number, they often are asked to look at two numbers and compare them, deciding which is more and which is less.

**holds, full, container** In Investigation 3, students compare containers on the basis of the number of cubes each can hold, an activity that involves counting and comparing numbers to 40.

## About Story Problems

For Investigation 4, students do a lot of work with story problems, both as a class and individually during Choice Time. These story problems are a very important part of the unit, and all students need a chance to work on them. To be sure the story problems are comprehensible to second-language learners, you may want to arrange for help from bilingual aides or parents to translate these problems, and if possible, ask them to be present in class during those sessions. If you do not have that support, take the time to make rebus drawings on the story problem cards as a comprehension aid.

## Multicultural Extensions for All Students

■ When the students are investigating different ways to show number on their fingers in Investigation 1, including some different ways used in Africa, demonstrate (or ask a visitor to demonstrate) how numbers are signed with American Sign Language. Two good sources for this information are *Signing for Kids* by Mickey Flodin (A Perigee Book, Putnam, 1991) and *My Signing Book of Numbers* by Patricia Bellan Gillen (Kendall Green; Gallaudet University Press, 1988).

■ If you are creating your own story problems, include problems about cultures that are familiar to your students. These problems may be about ethnic foods, clothing, the currency of other countries, or special cultural events.

# Investigations

# INVESTIGATION 1

# Visualizing Numbers

## What Happens

**Session 1: Quick Images** In an activity called Quick Images, students are briefly shown images, in this case dot patterns. After the image is removed, students draw or make a copy of what they saw, compare their copy with the original image, and share ways they thought about the image in order to remember it.

**Session 2: Compare Dots** Students play the game Compare Dots, in which they determine which of two dot patterns has more dots.

**Sessions 3 and 4: Copying Cubes** Students are given an object built from 10–15 interlocking cubes. They build a copy and find the number of cubes in the object. For the activity Copying Cubes, students build their own objects, trade with a partner, and copy each other's constructions. The remainder of these sessions is structured as Choice Time; students continue with Copying Cubes and Compare Dots.

**Sessions 5 and 6: Number Shapes** Students repeat the Quick Images activity. In a new activity, Counting Pattern Blocks, they create a picture, pattern, or design from about 20 pattern blocks. They record the total number of blocks they used and the number of each kind of block. The rest of the two sessions is Choice Time, with Counting Pattern Blocks, Copying Cubes, and Compare Dots.

**Sessions 7 and 8: Making Dot Pictures** Students do Quick Images with three dot patterns designed to demonstrate what characteristics of an image make it easy or difficult to find the total number of dots. Students then make their own Quick Image dot pictures, looking for ways to organize the dots so they are easy to count. During Choice Time, they work on this activity and Counting Pattern Blocks.

**Session 9: How Many Dots?** Students use their own Quick Image dot pictures to play a partner version of Quick Images. They explore different ways to show numbers on their fingers, and look at a book about finger-counting methods from Africa.

**Routines** Refer to the section About Classroom Routines (pp. 172–179) for suggestions on integrating into the school day regular practice of mathematical skills in counting, exploring data, and understanding time and changes.

## Mathematical Emphasis

- Counting quantities up to about 20
- Developing strategies for counting and comparing the number of dots in dot patterns
- Developing strategies for organizing collections of objects so that they are easy to count and combine
- Using numerals to record how many, for quantities up to about 20
- Representing quantities with pictures, with a variety of objects, and on fingers
- Exploring different ways to arrange a set of objects, such as rectangular arrays and equal-sized groups
- Using number combinations to describe different arrangements of a set of objects

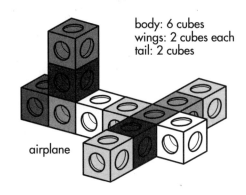

body: 6 cubes
wings: 2 cubes each
tail: 2 cubes

airplane

## What to Plan Ahead of Time

### Materials

- Overhead projector (Sessions 1 and 5–8)
- Interlocking cubes: at least 30 per student (Sessions 3–6)
- Pattern blocks: 1 bucket per 6–8 students (Sessions 5–8)
- *Count on Your Fingers African Style* by Claudia Zaslavsky (Black Butterfly Children's Press, 1996) (Session 9, optional)
- Small resealable plastic bags (25–40)
- Envelopes for storing transparencies
- Counters (such as buttons, bread tabs, or pennies): at least 40 per pair (available)
- Unlined paper (available for student use)
- Chart paper or newsprint (18 by 24 inches): 15–20 sheets (available for use as needed)

### Other Preparation

- Before Session 3, make a three-dimensional object from 10–15 cubes. For example, see the airplane on p. 2. All cubes must be visible so students can easily count them. Make an identical copy for each group of four students. Using cubes of mixed colors will help students count.
- Before Session 5, make a large display version of Student Sheet 3, Counting Pattern Blocks, on chart paper or newsprint.
- Before Session 9, make three or four dot pictures on a copy of Student Sheet 4, Making Quick Images. See p. 29 for ideas.

- Duplicate the following student sheets and teaching resources, located at the end of this unit. If you have Student Activity Booklets, copy only items marked with an asterisk.

#### For Session 1

Family letter* (p. 182): 1 per family (sign and date before duplicating)

Dot Cards, Sets A–D (pp. 187–190): 1 transparency of each card.* Cut apart. Keep each set in a separate envelope.

**Note:** Before making all the transparencies, test an image on the overhead. If it seems too small, enlarge on your copier, or draw larger on transparency film.

#### For Session 2

Dot Cards, Set A (p. 187): 1 set of 32 cards per pair, plus 1 per student, homework.

Dot Cards, Sets B–D*: 3 sets of each (If possible, copy on card stock and use a different color for each set.)

Student Sheet 1, Compare Dots (p. 183): 1 per student, homework

#### For Sessions 3 and 4

Student Sheet 2, Copying Counters (p. 184): 1 per student, homework

#### For Sessions 5 and 6

Student Sheet 3, Counting Pattern Blocks (p. 185): 1 per student

#### For Sessions 7 and 8

Student Sheet 4, Making Quick Images (p. 186): 1 per student and a few extras*

Easy and Hard Quick Images* (p. 191): 1 transparency. Cut apart.

# Quick Images

## What Happens

In an activity called Quick Images, students are briefly shown images, in this case dot patterns. After the image is removed, students draw or make a copy of what they saw, compare their copy with the original image, and share ways they thought about the image in order to remember it. Their work focuses on:

■ becoming familiar with different ways to arrange a set of objects, such as rectangular arrays and equal-sized groups

■ using number combinations to describe different arrangements of a set of objects

■ analyzing visual images

■ becoming familiar with combinations of numbers up to 10

■ describing position of and spatial relationships among objects

### Materials

■ Overhead projector
■ Dot card transparencies, Sets A and B
■ Family letter (1 per student)

## Activity

### Quick Images

The Quick Images activity is repeated several times in the grade 1 *Investigations* curriculum. As students do Quick Images throughout the year, they gain experience analyzing visual images and using number combinations and number relationships to describe different arrangements of a set of objects. They also become familiar with geometric arrangements, such as rectangular arrays.

**Note:** Quick Images is designed for use with an overhead projector, but if one is not available, you can present each of the images enlarged or drawn on letter-size sheets of paper.

Choose a transparency from Dot Cards, Set A, with either five or six dots. Place the transparency on the overhead projector, but do not turn it on yet. (Whether you orient the cards vertically or horizontally for Quick Images is not important.) Gather students where they will be able to see the image projected on the overhead. Distribute counters and pencil and paper to each student.

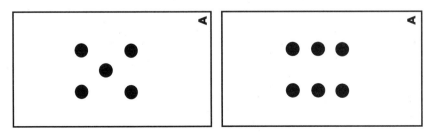

In the next few weeks, we'll be doing a lot of different things with numbers. We're going to start by doing an activity called Quick Images. Can anyone tell me what an image is?

If some students say that an "image" is "something that you imagine, something you see in your head," explain that in this activity that's what they'll be doing; they will be imagining a picture, or forming an image of a picture in their mind.

An image is another name for a picture. We'll be looking at some pictures, or images, of little black dots. We call it "Quick" Images because you'll only get to see the picture for a short time. Then I'll cover it up, and you will try to make a copy of the picture you saw.

Before turning on the overhead projector, explain where on the wall the dot image is to appear and remind them to look carefully because they won't have long to get the picture in their mind. To help students concentrate on the image, they should not draw anything or use counters while the image is visible.

OK. Hands in your laps. No pencils or counters when I'm showing the image. Here it comes!

Show the dot pattern for 5 seconds and then cover it.

Note: You may need to adjust the amount of time you flash the image. If you show the image for too long, you will see students simply copying the image from the screen, rather than building from their mental image; if you show it too briefly, they will not have time to form a mental image and will not be sure what to draw or build.

Students now make a copy of the dot pattern, using counters or pencil and paper. If some students are concerned that they cannot recall the figure exactly, assure them that they will have another chance to see the picture and to revise their work.

When students have completed their first attempts at representing the image, explain that you are going to show it again. Encourage students to study the picture carefully while it is visible.

Hands in your laps. Here comes the picture again. Look carefully!

Show the image for another 5 seconds. Then let students revise their arrangement of counters or create new drawings.

Ask for a few volunteers to tell you how many dots they saw in the picture, and how they know.

You might begin by reassuring students that you know it can be difficult to tell the number of dots when you've seen the picture only briefly. As students give their responses, encourage them to explain how they remembered the picture. If it helps them explain to the class, they can show their arrangements of counters on the overhead or copy their drawings on the board or a piece of chart paper. In the **Dialogue Box,** Seeing Dot Images (p. 7), students in one class describe how they thought of the images showing five and ten dots.

After this discussion, show the image again and leave it visible for further revision and for checking the number of dots shown.

Encourage students to talk about any revisions they made in their work. For example, students who drew dots may be willing to share their first drawings.

Repeat the activity, using another transparency from Dot Cards, Set A. If you began with the 6-dot card, you might next show the 4-dot or the 7-dot card. If you use related images, some students may begin relating visual and number patterns.

Continue the activity for the rest of the session, using transparencies from Dot Cards, Set B, which shows a different configuration for 3 to 10 dots. You might show the 5-dot, 6-dot, and 9-dot cards. Follow these steps each time:

1. Briefly show the image.

2. Ask students to make a copy of the dot pattern they saw, using counters or pencil and paper.

3. Show the picture again briefly.

4. Ask for a few volunteers to tell you how many dots they saw in the picture and how they know.

5. Leave the image visible so students can compare their copies with the actual image.

## Session 1 Follow-Up

**Homework**

**Family Connection** Send home the signed family letter or the *Investigations* at Home booklet to introduce your work in this unit.

## Seeing Dot Images

These students have been doing Quick Images with the 5-dot card from Dot Cards, Set A. They have finished their two attempts at making the image, and now the teacher is showing it so students can compare their work with the actual image. Here students are sharing the different ways they thought of the image; some broke the image into smaller parts, some "translated" the image into numbers, and some saw the image as a whole shape.

**How many dots did you see altogether?**

**Tony:** Five.

**Anyone else see a different number?** *[Pause]* **OK, then, what helped you remember what the image looked like?**

**Nadia:** Two dots on each side, and one more in the middle is five.

**Tony:** Two *[indicates the top row]* plus two *[the bottom row]* is four, and then one is more is five.

**Eva:** One diagonal, two diagonals.

**And how many in each diagonal?**

**Eva:** There's three in each. They cross in the middle, so it's five.

**Donte:** It's an X.

*Later in the session, students discuss how they thought of the image on the 10-dot card.*

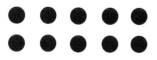

**How did you see this one? What helped you to make your own copy?**

**Chris:** I counted five on the top. And then I looked at both of them [both rows] and they were both the same, so I counted five and did five again.

**What does it equal, if you have five and five again?**

**Chris:** Ten.

**Kaneisha:** Kind of like I did, like Chris, but not that much. When I saw up at the top I saw three and two. Then I knew the bottom was the same so I did the same thing. Three plus two for both makes ten.

**Both Chris and Kaneisha saw the top row and the bottom row were the same and that really helped them.**

**Chanthou:** It has flowers in it. A lot of the pictures did.

**Can you show us what you mean?**

*Chanthou comes to the overhead, arranges four counters in a two-by-two array, and explains that this "flower" contains four dots.*

**How many flowers in this picture [the 10-dot image]?**

**Chanthou:** Two.

**How many left over?**

**Chanthou:** Two. So it's ten.

As students work with images containing more dots, they develop ways to break the image into manageable parts, for example, by identifying sections of the image that repeat or by seeing the image as made from component "objects." They also develop strategies for keeping track of and combining all the parts to find the total number of dots.

# Compare Dots

## Materials

- Dot Cards, Set A (1 set per pair, and 1 set per student for homework)
- Dot Cards, Sets B–D (3 sets each)
- Student Sheet 1 (1 per student, homework)

## What Happens

Students play the game Compare Dots, in which they determine which of two dot patterns has more dots. Their work focuses on:

- developing strategies for counting and comparing the number of dots in dot patterns
- using number combinations to describe different arrangements of a set of objects
- becoming familiar with combinations of numbers up to 10
- becoming familiar with different ways to arrange a set of objects, such as rectangular arrays and equal-sized groups

## Activity

### Teacher Checkpoint

### Compare Dots

If you have done the *Investigations* unit *Mathematical Thinking at Grade 1*, your students will be familiar with the game Compare, which involves comparing two number cards to determine which shows a larger number. In that case, students will probably need just a brief introduction to Compare Dots.

Introduce this game to the entire class by gathering students in a circle on the floor to watch a demonstration game. Either enlist two volunteers to play the game, or choose one student to play with you.

**Today we're going to play a game called Compare Dots. At the beginning of the game, each player gets half the cards in the deck.**

Demonstrate how to deal out the cards evenly between the two players.

**Players turn their cards facedown. Then, at the same time, they both turn up their top card. The player who has the card with more dots says "Me."**

After this has been done in the demonstration game, hold up the cards that each player turned over.

**Libby turned over this card [5 dots], and Nadia turned over this card [7 dots]. Which card has more dots? How do you know?**

Ask students to explain their strategies. Some students may count the dots on each card. Others may find the total by mentally grouping the dots on the cards in some way (for example, in 2's), and then combining the groups. Still others may compare the cards without finding the totals, noticing simply that one card has one more row of two dots.

Continue in this way through two or three more turns, or until you think students understand the game.

**Sometimes you might each turn up the same card. When that happens, both of you just turn over the next card. Then the player who has the card with the most dots says "Me."**

Explain that the game is over when players have turned over all their cards.

Pair up students to play the game for the rest of the session. Give each pair a deck of Dot Cards, Set A.

**Note:** Your students may be familiar with the card game War, a similar game. An important difference is that in Compare Dots, players do not win or lose because they do not capture each other's cards. Students who are familiar with War may naturally want to capture the other card when their number of dots is larger. You can decide how important it is for students to play less competitively; if the competition is distracting students from thinking about the numbers, you may want to insist that they play according to the Compare Dots rules.

## Observing the Students

Teacher Checkpoints are designated times for you to observe students at work. After students have begun playing in pairs, you can observe how they are counting up to 10 objects, comparing numbers up to 10, and using number combinations and number relationships to find the total number of dots on the cards. Circulate to observe and to offer support as needed.

- Can students deal out the cards evenly between two players? (You might suggest that the students repeat "one for you and one for me" while dealing, to help them remember how to distribute the cards.)
- How do students find the number of dots on a card? Do they count each dot? Do they group the dots in some way, such as by rows, and then combine the number in each group? Do they seem to "just know" how many dots are on each card? The **Teacher Note,** Finding the Number of Dots (p. 11), describes typical strategies students use and suggests how to help students explain their strategies.
- What strategies do students use for determining which card has more dots? Do they "just know" which number is larger?

Some students might try to decide by comparing the size and shape of the dot configurations, rather than by comparing numbers of dots. For example, they might say that a long, thin array of dots has more than a square array of dots. Ask these students to prove to you which card has more dots by finding the number on each card.

■ Do students play cooperatively? If you think students are playing too competitively, emphasize that getting the card with more dots is a matter of luck, not a matter of being a "better" player. Explain that good players play cooperatively: they check or help one another; they explain their thinking to one another; they ask one another for help; they wait while the other player takes the time to determine which card has more dots.

**Variations and More Challenge** You can use the Dot Cards, Sets B and C to provide a greater variety of dot patterns for each number. Some (but not all) students will find the Set A cards easiest to work with, because the arrangements are based on rectangular arrays. At some point now or during Choice Time (Sessions 3–4 and 5–6), introduce Sets B and C. You will need to determine when students are ready for these. If some pairs are just beginning to develop strategies for grouping dots on the Set A cards, wait before giving them a new set. On the other hand, if students have good strategies for grouping the dots in Set A, or if they no longer seem challenged, ask them to play with Set B or C. For extra challenge, students can play with Set D, which has 11–20 dots in rectangular arrays.

**Near the End of the Session** Alert students when 5 or 10 minutes remain in the session. Review cleanup procedures, including how to return materials to storage areas and how to double-check the floor for pencils, stray counters, and other materials.

## Session 2 Follow-Up

**Homework**

**Compare Dots** Students teach someone at home to play Compare Dots. Send home Student Sheet 1, Compare Dots (the game directions), and four sheets to make Dot Cards, Set A. If students are unlikely to have scissors at home, give them time during the school day to cut apart the cards. Encourage students to keep their cards in a special place at home, because they will be using them for other games at home over the next few weeks. Some teachers give students a Math at Home folder for storing the materials they will be bringing home throughout the year.

**Extension**

**Double Compare Dots** For a more difficult variation of Compare Dots, students turn over the top *two* cards in their piles on each turn. They determine the total number of dots on their two cards, and the player with the higher total says "Me."

# Finding the Number of Dots

As students play Compare Dots, observe them to find out more about their strategies for determining the number of dots on a card. Following are some strategies students typically use for finding the total number of dots, with suggestions for encouraging students to reflect on and communicate how they are thinking about the dots.

**Counting Dots** Often students count the dots on the card in order to determine or to check the total. Some students count by 1's, touching each dot as they count. Others count by the number of dots in a row, column, or group on the card (2 for the first row, 4 for the second, and so on).

When students first learn to say a particular skip-counting sequence ("2, 4, 6, 8 ..."), they may not yet recognize that each successive number in the sequence represents the addition of a particular quantity. Activities such as counting dots or other objects arranged in rows or groups of uniform size can help students connect the numbers in a skip-counting sequence to the quantities they represent. However, even as students become comfortable counting by 2's for small quantities, they may return to counting by 1's as quantities increase ("2, 4, 6, um ... 7, 8 . . . 9, 10"). Understanding of counting by numbers other than one develops gradually over the early elementary years.

Whether students are counting by 1's or by some other number, you can help them strengthen their understanding of quantity by encouraging them to recount the dots in a different way. Students counting by 1's can check their totals by recounting in a different order (for example, from right to left, instead of from left to right). Students counting by some other number can count again by 1's, or by a different number. This can help them begin to recognize relationships among different ways of counting.

**Breaking the Image into Parts** Some students see the dots on the cards as groups to be combined, rather than individual dots to be counted. They look for ways to break the group of dots into manageable parts, which they can think about separately, then combine. For example, one student explained how he found that there were 10 dots:

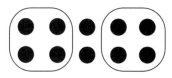

"Four [on the left] and four [on the right] is eight, and then two more [in the middle] is ten."

Another student explained that he got ten for the same card by breaking the group into two familiar patterns, a group of six and a group of four, declaring "and I know six and four is ten."

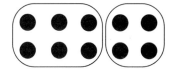

These students are beginning to recognize connections between dot patterns and number combinations: they see the total number of dots as the sum of smaller parts, not just as a collection of single dots. Being able to group numbers in different ways is an important part of developing effective strategies for adding and subtracting numbers. This kind of thinking develops gradually over the early elementary years. Encourage students who are beginning to think in this way to explain their strategies, but expect that they will still count by 1's to find or check the number of dots on many of the cards.

*Continued on next page*

**Using Equal Groups** Some students see certain dot patterns as sets of equal-sized groups (or, equal-sized groups with an extra part). When they first look at a card, they might note that it shows "three 3's," or "two 4's," or "two 2's and one more." Students might then combine the groups mentally ("3 and 3 is 6, and one more 3 is 9"), use some combination of mental addition and counting ("two 3's are 6, and then there's 7, 8, and 9") or, they might "just know" the total ("three 3's are 9").

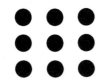

Learning to think about quantities as sets of equal-sized groups is important for situations involving multiplication or division. However, thinking in this way is challenging for most first graders. You will probably find that students take this approach only for a small number of groups of 2 or 3. Encourage students using this approach to explain their thinking, but don't expect them to see complex configurations as collections of equal-sized groups. Students' thinking about equal groups will continue to develop over the next few years as they have many opportunities to count and combine groups of different sizes.

**Recognizing Familiar Dot Arrangements** Some students may tell you that they know the number of dots on a particular card because they can tell from "just looking at it" or because they remember from previous work with the card. Even those who have committed a dot pattern to memory can benefit from explaining how it shows how a number can be broken into parts.

If you think a student has committed to memory a dot pattern with at least five or six dots in it, ask that student to look away from the image and describe it. Some students are quite skilled at memorizing images and then analyzing their mental images. When a student insisted that he "just knew" that a 3-by-3 array of dots showed 9, the teacher asked him to prove it, expecting him to do so by counting dots on the card. Instead, he looked away from the dot card and then pointed to each dot in an imaginary 3-by-3 array, saying a number for each "dot" he pointed to.

When another student was asked to explain how she "just knew" that a triangle shape showed six, she shut her eyes and explained that there were 3 dots on the bottom row, 2 in the middle row, bringing the total to 5, and 1 more on the top, so 6 in all.

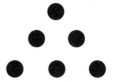

Analyzing mental images in this way can help students to develop a strong visual sense of the structure of a number—how it can be broken into parts—and to recognize the value of using mental imagery in solving mathematics problems.

# Copying Cubes

## What Happens

Students are given an object built from 10–15 interlocking cubes. They build a copy and find the number of cubes in the object. For the activity Copying Cubes, students build their own objects, trade with a partner, and copy each other's constructions. The remainder of these sessions is structured as Choice Time; students continue with Copying Cubes and Compare Dots. Their work focuses on:

- counting up to about 15 objects
- developing strategies for counting and combining collections of objects that vary by color and by arrangement
- developing strategies for counting and comparing the number of dots in dot patterns
- copying three-dimensional objects made from cubes
- becoming familiar with combinations of numbers up to 10
- becoming familiar with different ways to arrange a set of objects, such as rectangular arrays and equal-sized groups

### Materials

- Interlocking cubes (at least 30 per student)
- Identical objects built from 10–15 cubes (1 object per group of 4)
- Dot Cards, Set A (1 per pair)
- Student Sheet 2 (1 per student, homework)

## Activity

If your students have had frequent opportunities to explore interlocking cubes, they will be more ready to do this activity. If they have not used them recently, provide 10–15 minutes for free exploration with them first. You might also give students the chance to explore the cubes at other times during the day.

**Today you're going to copy something I've built out of cubes.** *[Hold up one of the cube objects.]* **I'll give one of these models to each group, along with a supply of cubes. Each one of you will take cubes and build an exact copy of this model. Your colors don't have to match mine, but you do have to make the exact same size and shape, using the exact same** *number* **of cubes.**

Students sit in groups of four so that they can share materials. Distribute the class set of interlocking cubes evenly to each group, and distribute the identical cube objects they are to copy. Each student builds a copy of the object. Students may touch the object if it helps them count the number of cubes, but they may not take the object apart.

### Copying Cubes

As students are working, circulate quickly to be sure they understand the task. As necessary, remind them to keep the model object in the center of the group so that everyone can see it.

When students have finished, ask for volunteers to tell how many cubes are in the object and to explain how they know. Some students will count one cube at a time, while others may begin to count in groups: "There are 6 cubes along the body of the airplane, and 2 more, 7, 8, for one wing, then 2 more for the other wing, that's 9, 10, and 2 more for the tail, that's 11, 12."

If some students disagree about the number, encourage them to recount to check the total. (If some disagreement remains, you need not take the time to resolve it.)

**Copying Cubes as a Choice** Explain that sometime over the next few days, students will work in pairs to build and copy other cube objects. Each student builds a model with at least 10 and no more than 15 cubes. Then partners exchange objects, and each copies the other's model. When they finish, they check that each copy is exactly the same size and shape as the original model, and that it uses the same number of cubes.

If it seems necessary, model the activity with a student:

**Let's say I built this airplane and Susanna built this flag. Then we switch, and Susanna builds a copy of my airplane, while I build a copy of her flag. When we're done, we have to agree on two things: We must agree that the copies are the same as the originals, and we must agree on the number of cubes in each object.**

Explain that students can touch the model object if it helps them to figure out how to build a copy, but they cannot take it apart. As in the whole-class activity, colors may vary.

**Note:** If you have established a rule that weapons are not permitted in the classroom, remind students not to build guns, swords, or similar objects.

---

## Activity

## Introducing Choice Time

If you are using the complete grade 1 sequence of *Investigations,* students will be familiar with Choice Time, when they choose among several activities to work on. The Choice Time format recurs throughout the *Investigations* curriculum. See the **Teacher Note,** About Choice Time (p. 166), for information about how to set it up and how first grade students can keep track of the choices they have completed.

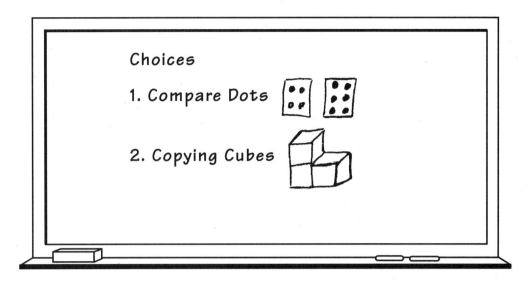

During each day of Choice Time, students choose one or two of the activities they want to participate in. They can select the same activity more than once, but should not choose the same activity every day. List these choices on the board or on a piece of chart paper. Include a simple picture with each choice as a visual reminder.

## Choice 1: Compare Dots

**Materials:** Dot Cards, Set A, 1 per pair; Dot Cards, Sets B, C, and D, 3 sets each.

Students are familiar with this game from their work in Session 2. Review the rules as necessary.

If you noticed in Session 2 that some pairs were just beginning to develop strategies for grouping the dots on the cards, suggest that they play again with Set A. Otherwise, ask students to play with Set B or Set C. Remind students not to mix the sets of cards.

For extra challenge, students can play with Dot Cards, Set D. By the end of Choice Time in Sessions 5–6, students should have played with at least two different Dot Card sets.

## Choice 2: Copying Cubes

**Materials:** Interlocking cubes (at least 60 per pair)

Students work with a partner. Each builds something with 10–15 interlocking cubes. Then they exchange and build copies of each other's objects. When both students have finished, they check that the copies are identical (except for color) and use the same number of cubes as the original.

Remind students to use no more than 15 cubes, as larger objects can be difficult to copy. Encourage them to keep all the cubes are visible in their model and avoid hidden interior cubes.

## Observing the Students

Observe and listen to students while they work on Choice Time activities. Recording your observations will help you keep track of how students are interacting with materials and solving problems. The **Teacher Note,** Keeping Track of Students' Work (p. 168), offers some helpful strategies. During this first Choice Time, watch for the following:

■ Do students try each choice, or do they stay with a familiar one? If, after a short time with one activity, students say they're done, ask them to tell you about what they have done and encourage them to investigate further.

■ How much do students interact with their partner? Do they share what they have done with others and observe what others are doing? Do they talk to themselves or others about what they are doing?

### Compare Dots

■ What strategies do students use to find the number of dots on a card?

■ What strategies do they use to determine which card has more dots?

### Copying Cubes

■ How do students keep track of the number of cubes they are using to create their objects? Do they recount the entire set from 1 each time they add a cube or two? Do they count up from a previous total? ("I had 10, and I just put on two more, so now I have 11, 12.")

- How accurately do students copy the number of blocks in each part of the object?

- How accurately do they copy the orientation of each component? That is, if a piece sticks out to the left in the original, does it stick out to the left in the copy?

- How do students count the cubes in the object? Do they count by 1's? Do they group the cubes in some way and then count the groups? ("Three for each arm, that's 6, then 4 for the head, that's 7, 8, 9, 10.") How do partners resolve any differences about the number of cubes in an object?

- How do students go about checking that a copy is identical to the original? Do they use counting strategies? Do they compare orientation of different parts? Do they compare size and shape? Do they ignore color differences?

Pairs having difficulty can work with a smaller number of cubes, or they can create simpler objects. For example, suggest they make something flat, or something with only three parts.

**Near the End of Each Session** Five or 10 minutes before the end of each Choice Time session, announce that it's time to stop working, put away the materials, and clean up the work area. If some students are in the middle of Copying Cubes, find a safe place for them to leave their cube objects until the next session. Students will have more time to work on both activities in Sessions 5–6. Once cleanup is complete, remind students to record the choices they completed, including which set of Dot Cards they used for Compare Dots. If you have posted lists, keep them for use again in other Choice Time sessions in this investigation.

## Sessions 3 and 4 Follow-Up

**Copying Counters** Send home Student Sheet 2, Copying Counters. This is a variation of Copying Cubes that students can do at home with common objects like coins, toothpicks, or paper clips. Each of two players makes a pattern or design from about 15 of these counters. Then, each player makes an exact copy of the other player's design. Players check that each copy uses the same number of counters as the original design and is exactly the same size and shape.

**Homework**

# Number Shapes

## Materials

- Overhead projector
- Dot card transparencies
- Interlocking cubes (at least 60 per pair, for about one third of the class)
- Pattern blocks (1 bucket per 6–8 students)
- Student Sheet 3 (1 per student)
- Large display copy of Counting Pattern Blocks
- Dot Cards, Sets A–D

## What Happens

Students repeat the Quick Images activity. In a new activity, Counting Pattern Blocks, they create a picture, pattern, or design from about 20 pattern blocks. They record the total number of blocks they used and the number of each kind of block. The rest of the two sessions is Choice Time, with Counting Pattern Blocks, Copying Cubes, and Compare Dots. Students' work focuses on:

- counting up to about 20 objects
- using numerals to record how many, including none (0)
- developing strategies for counting and combining collections of objects that vary by color and by arrangement
- developing strategies for counting and comparing the number of dots in dot patterns
- copying three-dimensional objects made from cubes
- becoming familiar with combinations of numbers up to 10
- becoming familiar with different ways to arrange a set of objects, such as rectangular arrays and equal-sized groups

## Activity

### Quick Images

Choose a dot card transparency that you have not yet used for Quick Images, such as the 7-dot card in Set B, and place it on the overhead projector. Gather students where they will be able to see the image projected. Distribute counters and pencil and paper to each student. Refer to Quick Images (p. 6) to review the steps of the activity.

**Today we're going to do Quick Images again. Who can remind us how we do Quick Images?**

Ask for a volunteer to explain the activity. If some students disagree, you might ask another student or two to tell what they remember. Keep this discussion brief and accept that students may remember the activity somewhat differently. If necessary, clarify that you will show a picture of some dots for a very short time, and then they will make a copy of what they saw.

Do the activity two or three times. To help students begin relating visual and number patterns, use related images. For example, if you first used the 7-dot card from Set B, next use the 8-dot card from the same set. Another related pair would be the 6-dot and 10-dot cards from Set C.

# Counting Pattern Blocks

If your students have not used pattern blocks recently, provide time for free exploration with them before introducing this activity.

Post your large display copy of Student Sheet 3, Counting Pattern Blocks, in the meeting area. Demonstrate the activity by quickly making a picture, pattern, or design from about 20 pattern blocks. Use all but one or two different types of pattern blocks, so that you can model filling in 0 somewhere on the sheet.

**I'm going to show you another activity you'll be doing in Choice Time. You start by making a pattern or design from about 20 pattern blocks. Here's the one I made.**

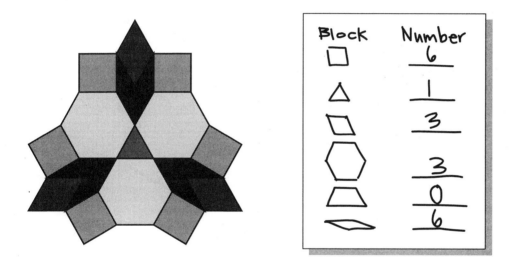

**Then, you count the number of each kind of pattern block you used. So, how many yellow hexagons do you see in the design I made? Does anyone see a different number?**

Even if someone has given the correct number, ask for another suggestion or two. This encourages students to verify their own solutions and not rely on you to tell them whether their answers are correct.

Ask volunteers to come up count each kind of pattern block in the design you have made, including those shapes you haven't used. Then, model filling in the number on the large recording sheet. Pay particular attention to using zero to record no blocks used.

**How many trapezoids did I use in my design? Can anyone find a trapezoid in my design? No? What do you think I should fill in here?**

Accept a few ideas. If no one suggests the use of 0, explain that we can use zero to show none of something. Model writing 0 on the sheet.

**Let's count the blocks to find out how many in all.... OK, we got 19. Is there a way we could check to make sure the count is correct?**

Some students may suggest recounting the blocks, perhaps in a different order. A few students may recognize that the total number of blocks is the sum of the numbers of the different kinds of block. Encourage them to explain their thinking, but accept that some students in the class may have difficulty following their reasoning.

Explain that as a Choice Time activity, each student will do just what you did: make their own design or pattern from about 20 pattern blocks, then find and record the numbers of blocks on Student Sheet 3.

## Choice Time

Post a list of the choices and briefly describe the options. Explain that students will be working on these choices today and tomorrow. By the end of math time tomorrow, they must have completed Copying Cubes, and they must have played Compare Dots with at least two different sets of Dot Cards. To help them remember what choices they have already completed, refer them to the records from the previous Choice Time (Sessions 3 and 4).

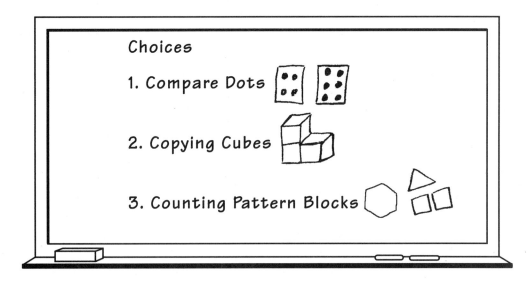

Often when a new choice is added to the list, there is a great deal of interest in trying the new activity first. Sometimes the quantity of materials available limits the number of students that can do an activity at one time. Even when this is not the case, limiting the number of students at each choice helps them make decisions about what they are going to do. It also encourages them to do some choices more than once. At the same time, support students in making decisions and plans for themselves rather than organizing them into groups and circulating the groups. Making choices, planning time, and taking responsibility for their own learning are important aspects of the school experience.

Set up the materials in three different locations in the classroom. You could use tables, clusters of desks, or rug space as places for students to work. Explain how many students can be at each center at one time. Depending on how you've organized your classroom, this could be indicated by the number of chairs at a certain table, or by posting the information on the board. For example:

1. Compare Dots              10 people

2. Copying Cubes             10 people

3. Counting Pattern Blocks   10 people

For a review of the Choice 1 and 2 activities, see Compare Dots and Copying Cubes (p. 15).

## Choice 3: Counting Pattern Blocks

**Materials:** Pattern blocks (1 bucket per 6–8 students); Student Sheet 3, Counting Pattern Blocks (1 per student)

Each student makes a design or pattern from about 20 pattern blocks, then finds and records the number of blocks used.

## Observing the Students

See pp. 16–17 to review what to look for while students are playing Compare Dots and Copying Cubes. For the new activity, watch for the following:

**Counting Pattern Blocks**
■ How do students keep track of the number of blocks they are using as they create their pictures? Do they recount the entire set from 1 each time they add a block or two? Do they count on from the last number they counted? ("I had 10, and I added 2 more, so now I have 11, 12.") Do they count accurately?

If some students are having difficulty counting up to 20 pattern blocks accurately, suggest that they work with fewer pattern blocks, perhaps 12. See the **Teacher Note,** Observing Students Counting (p. 23), for information on the range of first graders' counting abilities.

- Do students record the number of each type of block accurately? How do they show that there are none of a type of block? Do they use zero (0)?

- Do students use counting strategies to find the total number of pattern blocks? Do they count each block as they go in order around the picture? Do they count each type of block separately, perhaps counting on from each new type of block? ("There's 7 hexagons, then 4 triangles, that's 8, 9, 10, 11, then 2 trapezoids, that's 12, 13.") If their pictures contain repeated parts, do they count just one part and then repeatedly combine the numbers in that part? ("Each row of the pattern has 5 blocks, so 5 and 5 is 10, then 11, 12, 13, 14, 15 for three rows.") If some students tell you that they know the total because they kept track as they created their pictures, ask them to find a way to check.

- Do students combine the numbers on the recording sheet, rather than counting the pattern blocks? Some students may recognize that the sum of the numbers of each kind of block should equal the total number of blocks, but may be unsure of how to prove this. You might suggest they use a calculator to check. You might also ask them to find the total number of two or three kinds of blocks together, for example, the number of trapezoids and squares. They can then check by counting the blocks.

- What pattern block shapes do students know the names of? Do they recognize any relationships among them?

**At the Start of Session 6** Tell students that this is the last day that the choices Copying Cubes and Compare Dots will be available. Students who have not yet completed their work on Copying Cubes must do so today. Any students who have played Compare Dots only with Set A should spend some of their time today playing with Set B or C.

# Observing Students Counting

Your students will be counting many things during this unit and throughout the year. Counting involves more than knowing the number names, their sequence, and how to write them. It is the basis for understanding our number system and for almost all the number work primary grade students do.

In first grade, expect a great deal of diversity among your students. By the end of the year, many students will have learned the oral counting sequence up to 100 and will begin to recognize patterns in the sequence of numerals from 1 to 100. However, many first graders will not end the year with a grasp of quantities greater than 25 or so. Students develop their understanding of quantity through repeated experiences organizing and counting sets of objects. In first grade, many of the activities that focus on quantity can be adjusted so that students can work at a level of challenge that is appropriate for them. Early in first grade, some students will need repeated experiences with quantities up to 10, while others will be able to work with larger collections. Some students may be inconsistent—successful one time, and having difficulty the next.

Your students will have many opportunities to count and use numbers in this unit and throughout the year. You can learn a lot about what your students understand about counting by observing them as they work. Listen to students as they talk with each other. Observe them as they count objects and as they count orally and in writing. Ask them about their thinking as they work. You may observe some of the following:

- *Counting orally.* Generally students can count orally further than they can count objects or correctly write numbers. For some students, the oral counting sequence is just a song; they don't necessarily know that when they count one more, they are referring to a quantity that has one more. Students need many experiences counting and adding small quantities as they learn about the relationship between the counting words and the quantities they represent.

- *Counting quantities.* Some students may correctly count quantities above 20; others may not consistently count quantities smaller than 10. Some students may count the number of objects correctly when they are spread out in a line but may have difficulty organizing objects for counting themselves. They may need to develop techniques for keeping track of what they are counting.

- *Counting by writing numbers.* Many beginning first grade students are just gaining some competency in writing numerals. Young students frequently reverse numbers or digits. Often this is not a mathematical problem but simply a matter of experience. Throughout the year, students need many opportunities to see and practice the sequence of written numbers.

# Making Dot Pictures

## Materials

- Pattern blocks (1 bucket per 6–8 students)
- Overhead projector
- Easy and Hard Quick Images transparencies A, B, and C
- Student Sheet 4 (1 per student, plus extras)

## What Happens

Students do Quick Images with three dot patterns designed to demonstrate what characteristics of an image make it easy or difficult to find the total number of dots. Students then make their own Quick Image dot pictures, looking for ways to organize the dots so they are easy to count. During Choice Time, they work on this activity and Counting Pattern Blocks. Students' work focuses on:

- finding ways to organize sets of dots so they are easy to count
- counting up to about 20 objects
- using numerals to record how many, including none

## Activity

### Easy and Hard Quick Images

Gather students in an area of the classroom where they will be able to see the images projected on the overhead. Distribute counters and pencil and paper to each student. Place transparency A of the Easy and Hard Quick Images on the overhead projector, but do not turn it on yet.

**We're going to do Quick Images again. I'll show you a dot picture for a short time, and then you draw or build what you remember seeing. The Quick Images today are special—some are fairly hard. After you try them, we'll talk about what makes some pictures harder to remember than others.**

Proceed as usual for Quick Images, first with transparency A, then B, and finally C. (See p. 6 to review the procedure for Quick Images.)

As students are trying to remember and draw the individual images, they may spontaneously offer their ideas about why some images are harder or easier to work with. Remind students that they will have a chance to talk about this more after they have tried all three pictures.

**What's Hard and What's Easy?** Display all three Easy and Hard Quick Images transparencies together on the overhead.

**Did anyone find some of these pictures easier to remember than others? Which one was easiest for you? Why do you think so?**

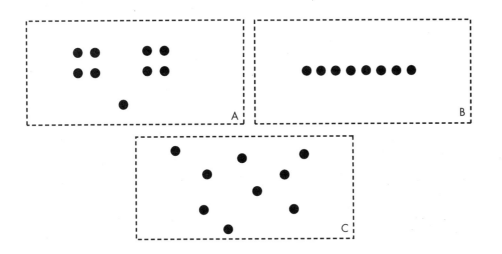

Encourage students to explain their thinking. Many students will say that image A was easiest because the dots are organized in rows that are easy to count; the dots are broken into small groups; and the dots are arranged in a familiar shape, a 2-by-2 square. Some students may say that they found it easiest to remember the shape of image B, but they found it difficult to actually count the dots.

**Which one was hardest to remember? Why do you think so?**

Some students may say that it was hard to count the dots on image B because they are so close together, and because they are not divided into smaller groups. Others will have found it difficult to keep track of the dots on image C. The **Dialogue Box,** Easy and Hard Quick Images (p. 28), shows how students in one class explained their ideas.

**Making Quick Images**  Explain that during Choice Time, one of the choices will be to create dot pictures that students will later use to play a game that is similar to Quick Images. Emphasize that they will be designing their pictures in a way that would help someone quickly tell how many dots are in them. In other words, they should try to make *easy* rather than *hard* images.

Show the class a copy of Student Sheet 4, Making Quick Images. Point out that students can make four dot pictures, one in each box. They can use 6, 7, 8, 9, or 10 dots. (You might assign particular numbers of dots to particular students, or you might let them choose.) At the top of the sheet, they record the number of dots they are using. They should use the same number in each box. Students may talk to their classmates about which arrangements they think will be easiest to work with.

## Choice Time

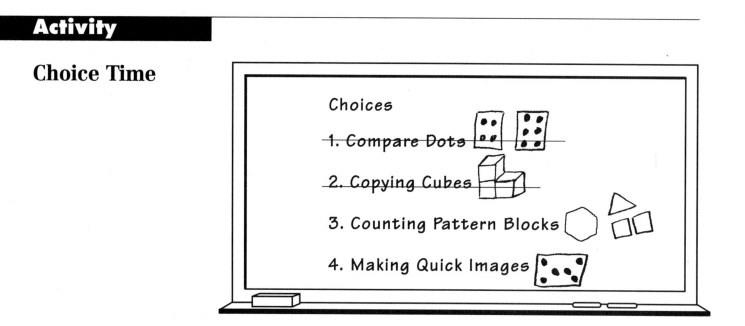

List the choices students will be working on for the rest of today and tomorrow. By the end of math time tomorrow, they must have completed both of the remaining choices.

To review Choice 3: Counting Pattern Blocks, see page 21.

### Choice 4: Making Quick Images

**Materials:** Student Sheet 4, Making Quick Images, 1 per student, plus extras; markers or crayons; scissors; counters

On Student Sheet 4, students create dot pictures that they will later use for How Many Dots? (a partner version of Quick Images). Remind them to design *easy* pictures—make them in a way that would help someone quickly see how many dots they contain. Students may want to model different arrangements with counters before actually drawing. They use the same number of dots in each picture and record that number at the top of the sheet. They may use crayons, markers, or pencils to make their pictures.

If you see some students creating dot patterns identical to those they worked with as Quick Images or on dot cards, ask them to think up new images.

Some students may need more than one copy of Student Sheet 4, Making Quick Images; hand out additional copies only as students request them.

When students finish, they choose the picture or pictures they think are easiest to count. They may ask classmates to help them decide. They cut out the box with each picture they have chosen and store in their math folders for use in the next session.

## Observing the Students

For more detail on what to observe as students work on Counting Pattern Blocks, see p. 21; the following questions are a brief reminder.

### Counting Pattern Blocks

- How do students keep track of the number of blocks they are using as they create their pictures?

- Do they record the numbers of blocks accurately? Do they use zero (0) to show no blocks of a particular type?

- What strategies do students use to find the total number of pattern blocks?

### Making Quick Images

- How do students arrange the dots? Do they organize them into smaller groups? into rows? Can they explain why they think their arrangement of dots would help someone count how many?

- Do students recognize that the way the dots are organized (not just the total number of dots) can make it easier or more difficult to count them? Encourage students to talk to a partner about which of their arrangements would be easiest to recognize for Quick Images.

- How do students count and keep track of the dots in their pictures? Do they count each dot individually? Do they find the number in different groups (in rows or clusters) and then combine the groups?

## Session 7 and 8 Follow-Up

**Double Compare Dots** Suggest that students return to the game of Compare Dots they took home earlier and play the variation Double Compare Dots with someone in their family.

 **Homework**

## Easy and Hard Quick Images

These students are talking about the Easy and Hard Quick Images.

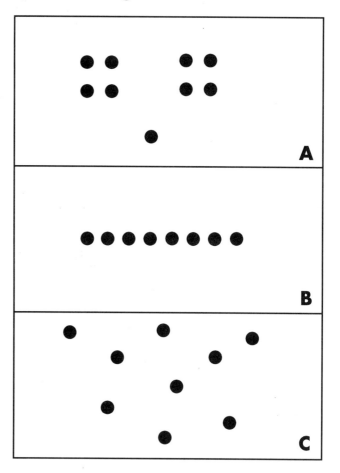

**Michelle:** The first (A) was easy, because it had 2 and 2, and 2 and 2.

**Luis:** It had two 4's and then 1.

**Eva:** The second (B) was easy to remember the shape, but hard because it was a line.

**Iris:** The last one (C) was hard. It was all scrambled out.

**Kaneisha:** I see a pattern! It went easy (A), hard (B), and harder (C).

**William:** The second was the hardest, because it was so straight.

**Kristi Ann:** The second one looked like it was 10, but it really wasn't.

**Luis:** The first was three different parts, so it was easier. It looked like 9, and it was 9!

**Garrett:** I think the third was really hard because it was going all different ways.

**Tamika:** It was all scrambled and not easy to count.

Although the students have different ideas about what made an image easy or hard, nearly *all* their ideas are based on whether or not an image can be broken into countable parts: A is easy because you can "see" the number of dots in each of several parts; B is hard because it's a straight line, so the dots are hard to count; C is hard because there is no obvious way to group and count the dots.

# How Many Dots?

## What Happens

Students use their own Quick Image dot pictures to play a partner version of Quick Images. They explore different ways to show numbers on their fingers, and look at a book about finger-counting methods from Africa. Students' work focuses on:

- using fingers to show combinations of numbers up to 10
- analyzing visual images
- using number combinations to describe different arrangements of a set of objects

### Materials

- Demonstration dot pictures
- Students' Quick Image dot pictures (from Sessions 7–8)
- Class sets of Dot Cards, A–C (available)
- *Count on Your Fingers African Style* (optional)

**Activity**

This game is similar to the Quick Images activity. One student briefly shows a dot picture to a partner, then hides it, and the other tries to recall the number of dots in the picture. Make counters and pencil and paper available for use as you introduce the game.

**Note:** Some of the dot images students create are difficult for others to copy. In order to keep the focus on developing strategies for finding the total number of dots, students are not required to make a copy of the image. However, they *may* do so if it helps them to recall the number of dots in the picture.

Demonstrate How Many Dots? by playing with a student volunteer. Give your volunteer the dot pictures you have made for demonstration. Ask the student to sit across from you and secretly choose a dot picture. If possible, sit so that when the student holds up the picture for you to see, the rest of the class will be able to see it, too.

## How Many Dots?

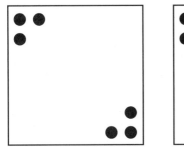 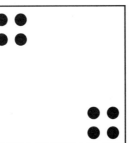 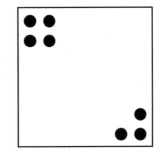

Sample of teacher's Quick Image
dot pictures for demonstration game

**Mia secretly chose a dot picture. She's going to show me the picture for 5 seconds, and I'm going to try to figure out how many dots are in the picture. She'll be counting to 5 to herself. For now, I'll help out by showing the count on my fingers.**

Demonstrate showing the count on your fingers: one finger for a silent count of 1, two fingers for a silent count of 2, and so on, up to 5 seconds.

**While Mia is showing me the picture, you look at it too. Try to see how many dots are in it.**

Tell your partner to announce "Here it comes!" before turning over the first picture. Once the picture is visible, begin the count to 5 seconds on your fingers. If necessary, remind your partner when it's time to hide the picture.

Encourage students to think about what they saw. If it helps students to remember, they can draw dots or build with counters, as they did with Quick Images. Allow for time for this, then ask for volunteers to tell the number of dots they saw and to explain how they remember.

Tell your partner when you're ready for the second showing. Again, the student should announce "Here it comes!" before showing the same picture for another silent count to 5.

**I'm ready to take a guess. First, who else knows the number of dots in the picture?**

After students have explained their ideas, make your guess at the number of dots. Then ask your partner to turn over the picture and hold it up for everyone to see.

Play once or twice more, as needed, until you think students can do the activity on their own.

Before students begin playing in pairs, practice counting out loud up to 5 at a moderate pace with the class two or three times. Then, have students practice silently two or three times. You might count along with them, mouthing the words for the numbers from 1 to 5, but not actually saying anything.

**Playing in Pairs** Each pair needs about 20 counters, paper, pencils, and the Quick Image dot pictures they created in Sessions 7–8. They should use only the picture or pictures they thought would be easiest to count. If some students do not want to use the dot pictures they created, they can use dot cards from one of the class sets.

To help establish an appropriate pace for counting to 5, you could start out with all the Player A's turning over their cards at once. Then you can silently count to 5 along with them, and remind them when to hide the pictures. Circulate to remind pairs of the next steps in the activity. As you observe the pairs, you may need to ask some to adjust the amount of time they show the image. For example, if students count too quickly, they may need to show the image for a count of 10.

Students may show one or more images each. Plan to spend no more than 20 minutes on this game in this session. Students who find this game especially engaging may want to keep playing outside of class time.

If you would like to use some of the students' own dot pictures for doing Quick Images with the whole class some time, duplicate their work on transparencies or enlarge them on letter-size paper.

# Numbers on Our Fingers

Different people have different ways to show numbers on their fingers. Introduce this activity by demonstrating one way to make a number with your fingers.

**When we solve problems, a lot of us use our fingers to show numbers and to count. Fingers are really useful ways of showing numbers. If I say I need this many pencils** [hold up the five fingers of one hand], **you'd all know how many I wanted, wouldn't you?**

Ask students for different ways to show numbers on their fingers.

**Everyone hold up three fingers.... Take a look around you. I see some different ways to show three. Who wants to tell about their way?...
So, Leah held up the middle three fingers on one hand. Did anyone else do something like that, three fingers on one hand? Who wants to tell about a different way?**

If students do not describe the number combinations that their fingers represent, model this for them:

**William held up two fingers—the pointer and middle finger—on his left hand, and one finger—the pointer—on his right. How do you know that shows three?**

**Who else showed three with two and one? Who did something else? So, Nathan held up the pinkie and thumb on one hand and the thumb on the other. Did anyone else do three with one and one and one?**

Ask students for ways to show several other numbers, such as 6, 7, and 8. See the **Dialogue Box,** Showing Six on Our Fingers (p. 35), for a discussion that took place in one class.

**African Finger Counting**  Take a few minutes to demonstrate a method of finger counting used in Africa.

**In some places, where people speak lots of different languages, they use finger signs to communicate about numbers with each other. Many different groups in Africa have ways to show numbers with their fingers.**

❖ **Tip for the Linguistically Diverse Classroom**  Point out Africa on a world map or a globe, and if possible, show pictures of this part of the world and the people who live there.

You can show students how to finger-count according to a method used in parts of Rwanda, Tanzania, and Cameroon. The gestures are as follows:

1  pointer finger extended on one hand

2  pointer finger and middle finger extended on one hand

3  three middle fingers extended on one hand

4  all fingers except thumb extended on one hand

5  a closed fist

6  three middle fingers extended on both hands

7  four fingers on one hand, three on the other

8  four fingers on each hand

9  four fingers on one hand, five on the other

[Source: Claudia Zaslavsky, *Africa Counts: Number and Pattern in African Culture*. Boston: Prindle, Weber & Schmidt, 1973. Paperback edition published by Lawrence Hill & Co., 1979.]

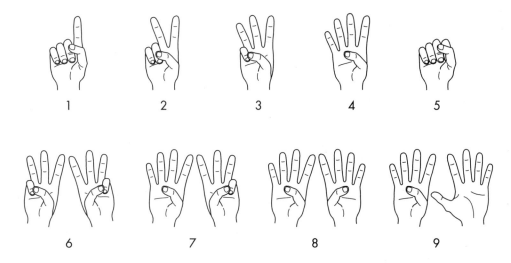

Have students practice the sequence with you a few times. Ask students for their ideas on what they notice about this way of counting:

**What does this way of showing 6 tell you about 6? What does this way of showing 9 tell you about 9?**

Note: The gesture for 10 is not given in Zaslavsky's book. You could ask students what they think it might be. A case could be made for either two closed fists or all 10 fingers extended, or even for a completely new gesture.

If you can get the wonderful children's book about finger counting, *Count on Your Fingers African Style* by Claudia Zaslavsky (Black Butterfly Children's Press, 1996), read it aloud to your class. The book shows ways to show numbers on fingers used in different parts of Africa, and includes numbers up to 20. Students can practice these ways, talk about what they show about number combinations, and then make up their own ways to show the numbers on their fingers for any numbers they have not yet worked with.

# D I A L O G U E ☐ B O X

## Showing Six on Our Fingers

For the activity Numbers on Our Fingers (p. 32), the teacher has demonstrated showing 5 by holding up all five fingers on one hand. Now the students are finding ways to show 6 with their fingers. As students share their ideas, the teacher encourages them to relate their ways of showing six to number combinations, and to begin thinking about similarities and differences among the various ways.

*[Several students show six by raising one hand and extending one finger of the other hand.]*

**Look, we have some different ways to show 6. What can you say about them?**

**Kristi Ann:** I did hand and ring finger.

**William:** I did the middle finger.

**Nathan:** They all did 5 and 1.

**Any other ways to show 6 with 5 on one hand and 1 on the other? Keep your hands up so everyone can see the ways we have so far.**

**Garrett:** Pointer.

**OK, Garrett showed one hand and the pointer. Anything else? OK, who has a way to mix up the numbers on your fingers, so it's not just 5 and 1?**

**Max:** Yeah, 2 here... no, 3. Then 3. *[He holds up the middle three fingers on each hand.]*

**Jamaar:** I can do another 3 and 3. *[He holds up both hands, then puts down the thumb and ring finger on each hand.]*

**How does that show 6?**

**Jamaar:** Because 2 on one hand, and 1. And then the same thing on the other hand.

**And how is that 6?**

**Jamaar:** Because 2 and 1 is 3. Um... and 3 on each hand, and 3 plus 3 is 6.

**Nadia:** I did something like that. I know 2 plus 2 equals 4. *[Holds up the thumb and pointer on both hands.]* Then I took 1... I did the same thing, so 5 *[raises one pinkie]*, 6 *[raises the other pinkie]*.

**So Jamaar thought 2 and 1 and 2 and 1 makes 6, and you thought 2 and 2 and 1 and 1 makes 6.**

**Kaneisha:** I did those fingers, but I started with 2 and 2, and I know that there's 3 left over.

**Three left over?**

**Kaneisha:** On each hand. And then I tried to... I had 2 plus 2, and I had 3 left over, and then I... took 1 from the 3. So it would be 3 and 3 on each hand.

**So, you ended up with the same fingers but you thought about it a different way. I see some of you have other ways, too. Who wants to tell about another way?**

**Tamika:** Put up 10. *[She holds up both hands.]* Count down 4. *[She puts down both thumbs and pinkies.]*

**Why did you count down 4?**

**Tamika:** Because 10, 9, 8, 7, 6.

**Eva:** I counted backwards too, but I used different fingers. I went 10 *[all ten fingers]*, 9 *[puts down right pinkie]*, 8 *[puts down right ring finger]*, 7 *[puts down right middle finger]*, 6 *[puts down right pointer]*.

**That looks like what some of you did earlier: 5 on one hand, and 1 on the other. But Eva thought of it as 10 minus 4, and some of you thought of it as 5 and 1.**

# INVESTIGATION 2

# Building Numbers in Different Ways

## What Happens

**Session 1: Twelve Cats and Dogs** In a How Many of Each? problem, students find combinations of cats and dogs they could have so there are 12 animals in all, and they share their solutions.

**Session 2: Problems About Ten** Students each make up their own How Many of Each? problem about ten. They solve their problems, record their solutions, and share their work with the class.

**Session 3: The Game of Double Compare** Students play the game Double Compare, in which they find which of two totals is larger.

**Sessions 4 and 5: Breaking Numbers into Two Parts** In On and Off, students toss a set of counters over a piece of paper and record how many land on and off the paper. In Counters in a Cup, one student secretly hides some of a set of counters in a cup, and a partner uses the remaining counters to determine how many have been hidden. These two games and Double Compare are offered during Choice Time. At the end of Session 5, students share the number combinations they got when playing On and Off.

**Sessions 6, 7, and 8: Towers of 10 and Number Choices** In the game Towers of 10, students use cubes of two colors to build towers 10 cubes high, then record the number of each color in each tower. Towers of 10 is added to the other games for Choice Time each day. Session 7 starts with more Quick Images, and at the end of Session 8, students share their combinations of 10 from playing Towers of 10.

**Session 9: Three Towers** As an assessment, individual students use cubes of two colors to make three different 10-cube towers. They record the number combinations represented by their towers. For the rest of the session, students solve a How Many of Each? problem about 15 things.

**Routines** Refer to the section About Classroom Routines (pp. 172–179) for suggestions on integrating into the school day regular practice of mathematical skills in counting, exploring data, and understanding time and changes.

## Mathematical Emphasis

- Finding combinations of numbers up to about 15
- Finding the total of two quantities up to 10
- Finding the larger of two quantities up to about 20
- Using pictures, stories, and objects to model number combinations
- Representing solutions to problems with pictures, numbers, and words
- Exploring relationships among combinations of a number

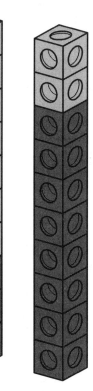

## What to Plan Ahead of Time

### Materials

- Number Cards: 1 deck per pair, manufactured or make your own (Sessions 3–5)

- Paper cups: 6–8 for the class (Sessions 4–8)

- Counters, such as buttons, bread tabs, or pennies: at least 30 per student (Sessions 4–8, and always available)

- Interlocking cubes: class set of 1000 (Sessions 6–9)

- Dot cubes: 12–16 for the class (Sessions 6–8)

- Overhead projector (Sessions 6–8)

- Dot card transparencies from Investigation 1 (Sessions 6–8)

- Chart paper or newsprint (18 by 24 inches): 15–20 sheets (available for use as needed)

- Unlined paper (available for student use)

### Other Preparation

- For Sessions 4 and 5, prepare large chart versions of the On and Off Game Grid on Student Sheet 8, and the Counters in a Cup Game Grid on Student Sheet 10.

- Sort your interlocking cubes by color (perhaps with student help) before Session 6. Keep each color in its own container for the rest of Investigation 2.

- Duplicate the following student sheets and teaching resources, located at the end of this unit. If you have Student Activity Booklets, copy only items marked with an asterisk.

*For Session 1*

Student Sheet 5, Turtles and Frogs (p. 192): 1 per student, homework

*For Session 3*

Number Cards (pp. 198–201): 1 set per student, homework. If you do not have manufactured cards, you will also need a class set for each pair; these will last longer if duplicated on card stock. Cut apart each set and store in a plastic resealable bag. For this investigation, wild cards are not needed and should be removed.

Student Sheet 6, Double Compare (p. 193): 1 per student, homework

*For Sessions 4–5*

Student Sheet 7, On and Off (p. 194): 1 per student, homework

Student Sheet 8, On and Off Game Grid (p. 195): 1 per student and a few extras*, plus 1 per student for homework

Student Sheet 9, Counters in a Cup (p. 196): 1 per student, homework

Student Sheet 10, Counters in a Cup Game Grid (p. 197): 1 per student and a few extras*, plus 1 student for homework

*For Sessions 6–8*

Game Record Sheet (p. 225): 1 per student, homework (optional)

# Twelve Cats and Dogs

## Materials

- Unlined paper
- Counters or cubes (available)
- Student Sheet 5 (1 per student, homework)

## What Happens

In a How Many of Each? problem, students find combinations of cats and dogs they could have so there are 12 animals in all, and they share their solutions. Students' work focuses on:

- finding combinations of 12
- using pictures, stories, and objects to model number combinations
- exploring relationships among combinations of a number
- recording solutions with pictures, numbers, and words
- finding more than one solution to a problem

---

## Activity

### Teacher Checkpoint

### Twelve Cats and Dogs

**Note:** How Many of Each? is a problem type that students encounter regularly in *Investigations* grade 1. Each time, they find combinations of two or more things that make up a given total. Over the year, they gain practice with number combinations, develop strategies for combining quantities, and find ways to record and organize their solutions. If your students did the unit *Mathematical Thinking at Grade 1,* they will be familiar with How Many of Each? problems and will probably need just a brief introduction.

**Introducing the Activity** Have a piece of chart paper and cubes or counters of two different colors at the meeting area. You might start out with a short discussion about pets at home. Ask students what pets they have, who has cats, who has dogs, who has both, how many of each type they have, and so on.

❖ **Tip for the Linguistically Diverse Classroom** Show pictures of cats and dogs. Sketch a house with a stick-figure family beside it. Sketch a cat and a dog and explain that these are the family's *pets.*

Explain that students are going to solve a problem about 12 pets. Then, present the problem that students will work on:

**Suppose I have 12 pets. Some of them are cats and some of them are dogs. How many of each could I have? How many cats? How many dogs? Remember, I have 12 things in all.** *[Write "12 in all" on the chart paper.]*

Accept two different student suggestions and model them with cubes or counters. Do not record the solutions on chart paper, as seeing the teacher use a particular recording method can make it difficult for students to develop their own methods.

If some students disagree, ask them to explain their thinking, but keep this discussion brief and keep the focus on explaining the task clearly enough so that students can find solutions on their own. Explain that there is more than one correct solution to this problem, and that students are to find solutions different from the ones that have just been suggested.

**Solving the Problem** Distribute unlined paper. Students work alone, in pairs, or in small groups. Encourage students to share their ideas with one another. They may use counters, such as cubes or buttons. When they have found a solution, they record it on blank paper using pictures, numbers, words, or a combination of these. Each solution should specify the number of cats and the number of dogs in either numbers or words.

Encourage students to seek more than one solution, but do not insist on it. See the **Teacher Note,** Exploring Multiple Solutions (p. 44), for a discussion of the variation in first graders' ability to understand that a problem may have more than one solution. A few of your students may be sufficiently challenged by finding just one solution. You can help them appreciate the variety of solutions by asking them to share their solutions with a partner; this will also happen when the whole class shares in the next activity. The **Teacher Note,** Finding Relationships Among Solutions (p. 45), discusses strategies that students in one class used to find more than one solution for this problem.

## Observing the Students

In this checkpoint, you have an opportunity to observe students' strategies for finding combinations of 12. Circulate to observe how they are approaching the problem and to offer support as needed.

- Do students understand what they are to find? Do they have a sense of how to begin? If any students are having difficulty making sense of the problem, ask them to put the problem in their own words. Talking through the problem may help them think of a way to approach it.

- How do students model the problem? Do they use counters? Do they write or draw anything to help them keep track of the number of cats and dogs? Do they work mentally?

  Students who are having difficulty getting started will probably find it helpful to use counters to model the problem. Suggest they begin by taking a few counters of one color. Then say:

  **Suppose those are the dogs. How could we find out how many cats there are?**

Or, you might put out 13 or 14 counters in any combination of two colors, and ask students to adjust the number of counters to make 12 in all. As you offer assistance, be sure not to show students a solution, as this can make it difficult for them to find one of their own.

- What strategies do students have for solving the problem? Some students might randomly take counters of two different colors, count them, and then add to or take away from the collection until they have 12 in all, perhaps recounting the set with each adjustment.

Other students may use a strategy that leads more directly to a combination of 12. Some might count out a few counters of one color, and then count up to 12 to find how many of the other color they need. Some might start with 12 counters of one color and then replace some with counters of the second color. Others might build on familiar number combinations: "I know 4 and 4 is 8, so I'll take 4 reds and 4 blues. Then I need some more, so I'll start at 8 and keep counting out blues until I get to 12." Some students may even start with known combinations of 12: "I know that 10 and 2 is 12, so I'll take 10 blues and 2 greens."

- Are students comfortable working with combinations of 12? Are they able to count 12 things accurately? If some students are having difficulty, suggest they work with a smaller number of cats and dogs, such as 8.

- How do students record their solutions on paper? Do they use pictures? squares or other symbols to represent counters they used to model the problem? numbers? equations? words?

  If some students record just the number of each animal (that is, "5 and 7" instead of "5 cats and 7 dogs"), ask them to find a way to show what each number represents: "How can we tell if that 5 means five cats or five dogs?"

- Do students find more than one solution? How do they do so? Do they find new solutions by changing something about a solution they already have, or do they treat each new solution as a separate problem to be solved? Do they compare new solutions with those already on their list?

- Do students check that their solutions are correct? What strategies do they have for checking?

Students who finish early might look for other solutions to the problem. They could also share their work with a partner.

**If Students Suggest Using Subtraction** Some of your students may show ways to make 12 with subtraction, for example, with an expression such as 14 – 2. Acknowledge that 14 – 2 is 12, and then encourage them to think about whether subtraction yields an appropriate solution to the problem at hand. For example, if a student recorded one solution involving subtraction and another involving a combination of 12, you might say:

**Your goal is to find a number of cats and a number of dogs that make up 12 in all. You wrote down 2 c 10 d. Does that solve the problem? What does it show?**

**You also wrote down 14 – 2. Does that show the number of cats and the number of dogs you need to make 12 in all? What does it show?**

Students may recognize that 14 – 2 shows that if there are 14 things and 2 are taken away, there are 12 left. While it is important that they think about why this does not fit the cats and dogs situation, do not be concerned if some students still think they can solve the problem with subtraction. In Investigation 4 of this unit and in later units at grades 1 and 2, they will work with many addition and subtraction situations and explore relationships and differences between them. For now, students interested in using subtraction to make 12 might work on the extension, Ways to Make 12 (p. 43). The **Dialogue Box,** Combinations of Seven (p. 67), demonstrates what happened in a different activity when a student suggested using subtraction to model an addition situation.

## Sharing Solutions

During the last 15–20 minutes of Session 1, bring students together, with their papers, to share some of their solutions.

**Who has a solution to share?... Chris got 4 dogs and 8 cats. Did anyone else get that? Did anyone get something different?**

Record the solutions on a piece of chart paper or on the board. Use a method of recording that you saw students using as they worked on the problem. For example, you might draw red squares for dogs and blue squares for cats, or use squares for dogs and circles for cats. Include the number of each kind of animal as well.

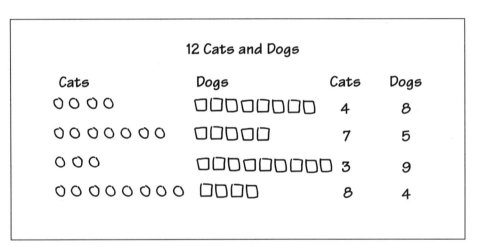

Each time a student suggests a solution, ask if it is the same as any you have already recorded. If the solution is already listed, acknowledge that it is a correct solution, but do not record it again. By asking students to attend to the solutions you have already recorded, you help them to appreciate the importance of keeping track. You may also help them notice relationships among different solutions, for example, opposite pairs (8 cats and 4 dogs; 4 dogs and 8 cats), or solutions in which the numbers differ by 1 (10 cats and 2 dogs; 9 cats and 3 dogs).

If everyone is eager to continue sharing, continue accepting and recording solutions. However, after several minutes, some students may find it difficult to remain engaged and focused. Before ending the discussion, acknowledge that there may be still other solutions. For those students still eager to share, you might set up a time later in the day when they can add their solutions to the chart with a partner or in a smaller group.

# Session 1 Follow-Up

**Turtles and Frogs** Students solve another How Many of Each? problem on Student Sheet 5 with a total of 11 turtles and frogs. Consider adjusting the number on this sheet; depending on how difficult students found the Twelve Cats and Dogs problem, you might stay with a number at about the same level of difficulty (10 or 13), or suggest either a smaller total (8 or 9) or a more challenging number (17).

A smaller total (8 or 9) is good for students who are just beginning to use their knowledge of number combinations to solve these problems, or who seem ready to do so. With a larger total, such students are likely to fall back on counting by 1's.

**Finding All the Solutions** Students try to find all the solutions to the Twelve Cats and Dogs problem. They explain how they know they found them all.

**Cats, Dogs, and Rabbits** Students again have 12 animals, but three different kinds: cats, dogs, and rabbits. They determine how many of each kind of animal they could have.

**Ways to Make 12** Students find and record different ways to make 12 (such as $9 + 1 + 2 = 12$ or $14 - 2 = 12$). They may use calculators.

 **Homework**

**Extensions**

Some of the problems in this unit (and throughout the *Investigations* curriculum) have more than one solution. Working on problems with multiple solutions opens up possibilities for clarifying one's thinking and for exploring a variety of mathematical relationships: *Have I found all the solutions? Is there another way of thinking about the problems that would yield more solutions? Are there any relationships among the solutions I found?*

Exploring questions such as these can be very exciting for some students, but initially quite challenging for others. Many of us—at all levels of mathematical knowledge—are more familiar and comfortable with problems that have a single correct answer.

In the activity Twelve Cats and Dogs (p. 38), students solve a problem that has many solutions. As you watch students working on this problem, you may notice a great deal of variation in their ability to understand that some problems have more than one solution. Some may discover on their own that there is more than one solution. Others will realize this when they share their solutions with others.

Still others may not be ready developmentally to understand that some problems can have more than one solution. They may accept that students in the class have come up with different solutions, but they may not be ready to look for multiple solutions on their own. Or, they may believe that only one of the multiple solutions can be correct.

As your class works on problems with multiple solutions throughout the year, continue to provide opportunities for students to share their solutions with others. As individual students become ready, encourage them to look for multiple solutions and to explore relationships among solutions, as described in the following Teacher Note. You will find that over time, more and more of your students will begin seeking multiple solutions on their own and exploring relationships among them.

# *Finding Relationships Among Solutions*

Like other How Many of Each? problems, Twelve Cats and Dogs is a complex situation that requires coordinating and keeping track of three pieces of information: the total number of cats and dogs, the number of cats, and the number of dogs. In order to solve the problem, students must count and keep track of a set of objects while comparing the number accumulated so far to the required total: *Do I have 12? Do I need more? fewer?* At the same time, they need to keep in mind how the two parts combine to reach that total: *I have 14. If I take away 2 things so that I only have 12, now how many cats will I have? How many dogs? How can I change what I have to get a different combination?*

While we as adults can see that the solutions to this problem are all the combinations of 12 with two addends, your students are unlikely to think of the problem in this way. Early in first grade, students who find several solutions are likely to work almost randomly, not noticing relationships among solutions, and not using one solution to find another. Although they keep a record of their work, they may not notice when they arrive at the same solution more than once. Some of these students may even tell you they have found all possible solutions—not because they have ways to check, but because they have so many solutions, they think there couldn't possibly be more. Over the year, as these students learn more about number combinations, they will begin to develop their own ways of working more strategically.

A few of your students may already see some relationships among the solutions they have found. They may find ways to change one solution to get others, perhaps by varying addends systematically (1 cat 11 dogs, 2 cats 10 dogs, and so on). Some may use their knowledge of relationships among number combinations: "I got 6 cats 6 dogs. I know 3 and 3 is 6, so if I take 3 cats and make them dogs, I get 3 cats and 9 dogs."

As you observe students working on this problem, you can learn a lot about how they are thinking about number combinations and the relationships among them. You will also see how they use what they know about number combinations to solve a complex problem. The examples below are drawn from a class in which many students found several solutions to the problem. Many of these students took an approach similar to Kristi Ann's (described first): They found several solutions using counting and perhaps some knowledge of number combinations, but did not seem to notice relationships among them. We then describe the work of two students in the class who are beginning to recognize relationships among number combinations.

### Finding Several "Unrelated" Solutions

Soon after the class begins working, the teacher visits Kristi Ann. The girl explains that she has three cats, so she started by building a train of three black cubes to represent cats. Now she's adding green cubes for dogs. After adding five green cubes, Kristi Ann quickly counts the whole cube train and gets eight. She adds three more green cubes. This time, she counts more slowly, and gets 11. She adds another green cube, says "12," and then carefully recounts the whole stack to check. She records 3 for cats, and counts the green cubes in the train to determine the number of dogs. She announces that there are nine and records this.

Kristi Ann explains that next, she's going to use two greens to show her grandmother's two dogs. She connects two green cubes and begins adding blacks, stopping occasionally to count the total number of cubes.

*Continued on next page*

When the teacher returns later in the session, Kristi Ann has recorded ten solutions, including two repeated solutions. She explains that she started out with the real numbers of pets belonging to people she knew, but then just decided to choose a number, and then find how many more she needed to make 12. When the teacher asks if there are more solutions, Kristi Ann seems uncertain, and does not seem to have a way to find out.

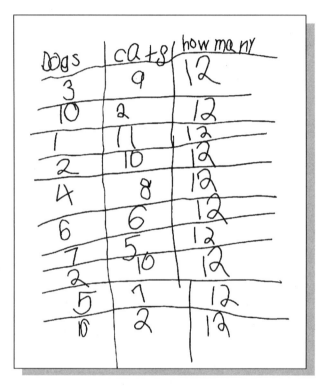

The problem is at an appropriate level of challenge for Kristi Ann. She is counting up to 12 accurately, is able to keep all the parts of the problem in mind, and has strategies for checking her work. Nonetheless, the teacher decides she should work on a How Many of Each? problem with a slightly smaller total (9) for homework. Although Kristi Ann relies on counting strategies to find combinations of 12, the teacher knows she is familiar with some combinations of smaller numbers. With a smaller total, Kristi Ann may be able to use her knowledge of number combinations to find solutions, and may begin noticing some relationships among the solutions she has found.

## Breaking 12 into Parts and Recombining the Parts

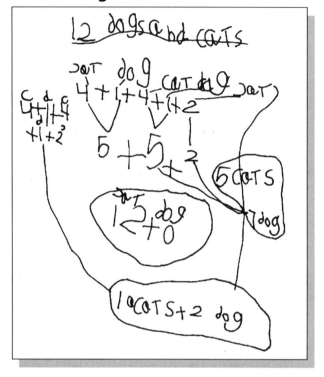

When the teacher arrives, Luis is puzzling over what he has recorded. He explains that he found solutions by first breaking 12 into 4 + 1 to make 5, another 4 + 1 to make 5, and then 2. He started by writing these two ways of breaking up 12 in two lines on his paper:

$$4 + 1 + 4 + 1 + 2$$
$$5 \quad + \quad 5 \quad + 2$$

By adding all the numbers up again, he got 12 cats and 0 dogs. By recombining the numbers in his first expression (4 + 4 + 2, and 1 + 1), he got 10 cats and 2 dogs. By recombining a 5 and the 2 in his second expression, he got 5 cats and 7 dogs. He tried making every alternate number in the first line a cat (see his writing at the upper left), but ended up with 10 cats and 2 dogs again. Now, he's trying to find another way to combine the numbers to make something different.

The teacher knows that Luis is familiar with many number combinations. However, organizing work carefully and working systematically is hard for him, as it is for most first graders. She briefly considers asking him to recopy his work and find other ways to combine the addends in the expression 4 + 1 + 4 + 1 + 2, but decides instead to encourage him to find more solutions by building on what he knows about number combinations. She hands Luis a clean sheet of paper and asks him to think of new ways to break 12 into parts.

**So, when you started out, you found one way to show 12. You did 4 and 1 and 4 and 1, and then 2 more. Is there another way you could break 12 into parts? Other numbers you could add to make 12?**

Luis stares at his paper a moment or two, and then says "Oh! I could do 2's." He records 2 + 2 + 2 + 2 + 2, notes that he's up to 10, and then adds + 2. He then circles the first three of the 2's, and says he could do 6 + 6. Then, he says, he could try just two of the 2's.

The teacher leaves to visit another student, but intends to return soon to see if Luis has found other ways to combine the 2's. She will suggest he compare solutions with Yukiko, who broke 12 into four 3's, and found all the ways to combine the 3's into two addends. She also makes a mental note to encourage Luis to organize his work more carefully when he is working on simpler problems.

### A Systematic Approach

Near the end of class, Yanni announces that he has found all the solutions.

**How do you know you have them all?**

**Yanni:** I started with 6 cats and 6 dogs because I know 6 and 6 is 12. You can take one cat and make it a dog, and it's 7 cats and 5 dogs, then it goes 8, 9, 10, 11. Then I did it the other way around, and did the 0's at the end.

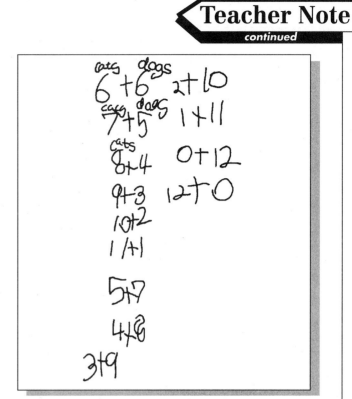

The teacher points to one of the combinations, 8 and 4, and asks Yanni how he knows it equals 12. He counts "9, 10, 11, 12," keeping track of each number on his fingers. She points to another combination, 3 and 9. He begins counting on from 3, but loses track of his count and ends up at 11. The teacher asks him to count again, more slowly. This time, he ends up at 12. The teacher asks about several of the other combinations Yanni has recorded. Some he "just knows," some he quickly determines by counting up from one of the numbers, and others—in particular, those that begin with a smaller addend—he has some difficulty with.

Although Yanni is not fluent with all the number combinations he has recorded, he is able to reason about relationships among number combinations and has a good understanding of why his strategy works: If you increase the number of cats by 1 and decrease the number of dogs by 1, you still have 12. To give Yanni more practice with number combinations and a chance to think more about ways to make 12, she asks him (for homework) to solve the problem with *three* kinds of animals and a total of 12 in all.

# Problems About Ten

## Materials

- Unlined paper
- Counters or cubes (available)

## What Happens

Students each make up their own How Many of Each? problem about ten. They solve their problems, record their solutions, and share their work with the class. Their work focuses on:

- finding combinations of 10
- making up and solving a story problem
- recording solutions with pictures, numbers, and words
- finding more than one solution to a problem

---

## How Many of Each? Problems About Ten

Distribute paper to students for making up and solving their own How Many of Each? problems.

**Yesterday, you solved a problem about 12 cats and dogs. Today, you'll be making up your own problem to solve. In your problem, there are 10 things in all. There are two different kinds of things—you get to decide what they are. They could be rabbits and birds in the woods, they could be oranges and apples in a fruit bowl. Or they could be any two things you like.**

**First, decide on what two kinds of things there are. Then, decide how many of each there are, so that the total number is 10.**

If some students had a great deal of difficulty working with 12 things for Twelve Cats and Dogs, you might suggest a smaller number for them, but most students should be working with a total of 10.

Students work alone, but they may share ideas with one another. They may use counters, cubes, or other materials. Circulate quickly to make sure students understand the task. If some students are having difficulty thinking up a problem, suggest they talk to classmates about possibilities. Encourage them to think up their own problem contexts, rather than using ones they have already worked with (such as dogs and cats). If you think it would be helpful, you might call the class together briefly to brainstorm possible items. Keep the discussion focused on general categories of things (such as kinds of pets, kinds of food, clothing in a dresser, toys in a toy box) rather than specific items (such as dogs and cats, or oranges and lemons).

Students' solutions should include three pieces of information: (1) the number of each thing, (2) the kind of each thing, and (3) the total number of things (ten). For example, any of the following would be acceptable ways of recording work:

words

pictures and numbers

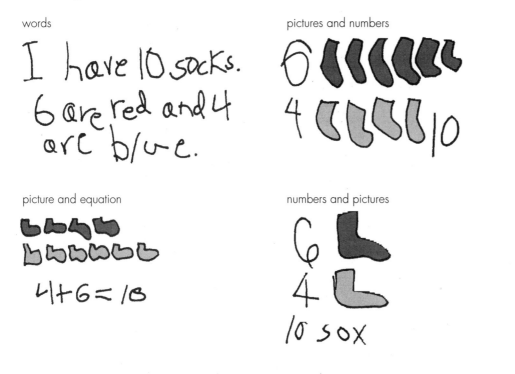

picture and equation

numbers and pictures

## Observing the Students

Continue to circulate to observe how students are approaching the problem and to offer support as needed.

■ What strategies do students have for solving the problem? Do they have strategies that involve combinations of 10? Do they solve the problem by collecting and counting a set of objects and adjusting the number until they have 10?

■ How do students model the problem? Do they use counters? pictures? numerals? Do they work mentally?

■ What strategies do students have for checking that their solutions are correct?

■ How do students record their solutions on paper? Do they use pictures or symbols? numbers? equations? words? If necessary, remind students that they need to show what the problem is about and the total, not just the number of each kind of thing.

- Do students find more than one solution? Do they have strategies for using one solution to find others, or do they seem to find new solutions at random?

If some students find and record a solution before most of the others are finished, they can share their problems and solutions with a partner. They can also look for additional solutions to their problem.

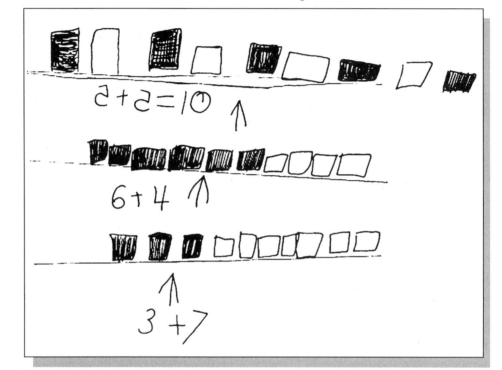

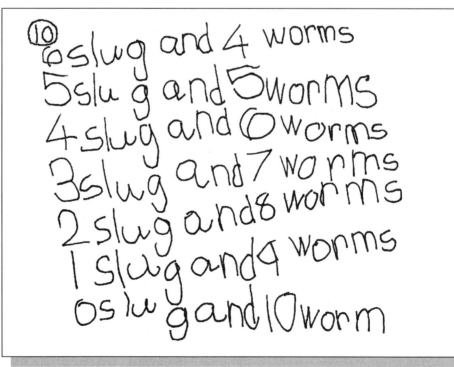

## Sharing Our Problems About Ten

During the last 20 minutes of the session, gather students to share their work. (Students will need to bring their papers.)

First, take about 5 minutes to focus on the variety of different objects they used in their problems. As students share, do not discourage them from telling their solutions, but keep the focus on what their problems were about.

**You made up a lot of different problems about ten things. Who would like to share theirs with the class? … Fernando made up a problem about red socks and blue socks. Who made up a problem about something else?**

To help your students appreciate the variety of contexts that can be used for story problems, you might call on students whom you know have made up problems about different kinds of things.

❖ **Tip for the Linguistically Diverse Classroom** Help second-language students find the words to identify the items they have drawn or used in their problems.

Take another 5–10 minutes for students to share how they recorded their work. Again, as students share, do not discourage them from telling their solutions, but keep the focus on how they recorded. It is important for students to have this exposure to different recording methods and the chance to explain their recording methods to others.

**I saw that different students recorded their work in different ways. Some of you used pictures, some of you used numbers, and some of you used words. Who wants to share how you recorded?**

Ask volunteers to hold up their work and explain how they showed what the problem is about, the number of each type of thing, and the total number of things. Since you will not have time to hear from everyone in the class, ask others to raise their hands if they recorded in the same way. That way, you can acknowledge the work of all students.

❖ **Tip for the Linguistically Diverse Classroom** Students who are not fluent in English may point to the elements of their work. Ask questions that can be answered with one-word responses. For example, "Did you show your total with a *number* or a *picture*?"

**Andre showed his solution with numbers and pictures. Did anyone else show their solution with numbers and pictures? Who showed their solution in a different way?**

Call on students who have recorded in different ways: one whose recording methods included numbers, one who used pictures and numbers, and one who used words.

If no one has used equations, do *not* introduce them yet. You will introduce them in Investigation 4, and students will have opportunities to use equation there and in the later grade 1 unit *Number Games and Story Problems*.

However, *if* some students recorded with equations, ask them to share their work. Encourage them to explain their recording methods, including the meaning of the + and = signs. If necessary, help clarify the meaning.

**Tuan's problem is about lemons and oranges. He showed his solution by writing 2 + 8 = 10.** *[Write this on the board.]* **For Tuan's problem, the first number, 2, shows how many lemons. The second number, 8, is the number of oranges. This addition equation shows that if you add them together [indicate the + sign], there are 10 things in all** *[indicate the = sign].* **You'll be seeing more addition equations like these a little later this year.**

## Session 2 Follow-Up

 **Extensions**

**Finding All the Solutions** Students try to find all the solutions to the problems they made up. They explain how they know they found them all.

**Combining Solutions** Working in pairs, students solve a How Many of Each? problem about ten things. Then they combine their solutions to find out how many of each and how many in all. For example, suppose Shavonne and Jonah solve this problem:

I have 10 toys in my toy box. Some are cars and some are dinosaurs. How many of each could I have?

Shavonne's solution is 6 cars and 4 dinosaurs. Jonah's solution is 3 cars and 7 dinosaurs. They find that together, they have 9 cars and 11 dinosaurs, or 20 in all.

**Larger Numbers** Give students a How Many of Each? problem about a large number of things. For example:

I have a string of 40 beads. Some are silver and some are gold. How many of each could I have?

They may use calculators to find their solutions.

# The Game of Double Compare

## What Happens

Students play the game Double Compare, in which they find which of two totals is larger. Their work focuses on:

- finding the total of two quantities up to 10
- finding the larger of two quantities up to 20

### Materials

- Number Cards (1 deck per pair, and 1 set per student, homework)
- Student Sheet 6 (1 per student, homework)

## Double Compare

Double Compare is a number card game similar to Compare Dots. The 44 Number Cards are dealt evenly to two players, facedown. Each player turns over the top two cards and finds the total of those two numbers. Then, players compare totals to determine which is larger.

If you are teaching the full grade 1 sequence of *Investigations,* students played this game in the first unit, *Mathematical Thinking at Grade 1,* and will need only a brief introduction. Play a demonstration game to review or introduce the rules. Enlist two student volunteers to play, or choose a student to play with you. Have counters available for students to use to figure out the totals.

Explain that the game is similar to Compare Dots, but players use Number Cards instead of dot cards, and each player turns over two cards at a time.

**When you have turned over the top two cards, find the total of those numbers. Then look at the other player's cards. The player with the larger total says "Me."**

Deal the cards and start the demonstration game. If you think that some students might have trouble seeing the cards that have been turned over, quickly sketch them on the board or chart paper. Include both the numbers and the objects (although you might draw circles in place of the objects).

**Fernando turned over 2 and 9. What's the total? How do you know?**

Give the class time to find the total of 2 and 9, and then ask volunteers to explain how they found the total. Students should explain their thinking whether it is correct or incorrect. Encourage students to share several different approaches.

Some students might have counted all the objects on the cards. Some may have used counters, and others counted up 2 from 9. Turn the students' attention to the other player's cards.

**Mia turned over 5 and 5. What's her total? How do you know?**

**Fernando's total is 11, and Mia's is 10. Which is larger? How do you know?**

If players have the same total, they simply turn over the next two cards, find their total, and compare as usual. The game is over when players have turned over all their cards.

For the remainder of the session, students pair up to play Double Compare. Each pair needs a deck of Number Cards with wild cards removed.

You might introduce the following variation to some pairs after they have already played a few games: The player with the smaller total says "Me."

## Observing the Students

Circulate as students play the game and watch for the following:

- Can students read and interpret the numerals on the cards, or do they count the objects on the cards to figure out the number? Can they count the objects accurately? If anyone is having difficulty distinguishing 6 from 9, they can count the objects on the cards to check, and always orient the cards so the number is on top.

- How do students combine the two quantities? Do they count all the objects? Do they count up from one of the numbers? Do they know any of the sums? which ones?

- What strategies do students use for determining which total is larger? Do they "just know" which number is larger? Do they count objects on the card to help them? Do they use counters? Do any students reason about which total is larger without actually finding the total? ("I know 5 + 9 is larger than 6 + 3 because 9 is more than 6 and 5 is more than 3.")

- Do students play cooperatively? If students are playing too competitively, emphasize that getting the larger card is a matter of luck, not a matter of being a "better" player. Remind them that good players check or help each other, explain their thinking, ask each other for help, and wait patiently while a partner finds a total or determines which number is larger.

**Different Levels of Challenge** Students having difficulty can play a simpler version of Double Compare called Compare, in which players take only one card on each turn. The student with the card that shows the larger number says "Me." The **Teacher Note,** Double Compare: Strategies for Combining and Comparing (p. 56), describes some of the many strategies students use to play the game and ways to adjust the game to make it more appropriate for students at different levels. See the **Teacher Note,** First Graders: A Wide Range of Understanding (p. 169), for information about adjusting activities throughout the year for the wide range of mathematical understanding in your class.

## Session 3 Follow-Up

**Double Compare** Students teach someone at home to play Double Compare. They will need the directions on Student Sheet 6 and a set of Number Cards. If students are unlikely to have scissors at home, give them time at school to cut apart their cards. You may want to provide envelopes to hold each set. (If you have done the unit *Mathematical Thinking at Grade 1,* students should already have a set of Number Cards at home.)

**Triple Compare** This game is played like Double Compare, except that each player turns over *three* cards on a turn and finds the total. The player with the larger total says "Me."

# Double Compare: Strategies for Combining and Comparing

Through the game of Double Compare, students develop their strategies for combining two numbers and for reasoning about quantity. The following scenes from a classroom illustrate situations that commonly arise, and show how to adjust the game for students at different levels.

### Counting Objects

As Garrett turns up 3 and 0 and Shavonne turns up two 8 cards, Garrett begins by reminding himself, "Count those little things [the pictures on the cards]." Then, while Shavonne watches, he counts each picture on the 3 card, touching them as he says the numbers. He announces that he has 3.

Shavonne places her cards side by side, overlapping the edges. She counts slowly, touching all the pictures as she says the numbers, but she skips a few pictures, counts a few twice, and comes up with a total of 13. Garrett says that he thinks 8 and 8 is 18. Although they are aware that at least one of these totals is inaccurate, they realize that regardless, Shavonne's total is greater than Garrett's total of 3, and they are ready to move on.

At this point the teacher steps in and asks them to recount Shavonne's total, slowly. After a couple of trials, Shavonne and Garrett both come up with a total of 16. The teacher suggests that they use interlocking cubes to help them find the totals on their cards. Since cubes, unlike the pictures on the cards, can be moved around, they can make it easier for students to keep track of what they have counted and what they have left to count.

Shavonne and Garrett both need to count by ones to be sure of their totals, and counting totals greater than 10 is challenging for them. The teacher plans to return in a few minutes to see if the cubes are helpful. If Shavonne and Garrett are still having difficulty working with larger numbers, she will suggest that they play with only the 1–6 cards. Later in the session, she will call together students having difficulty and will work with them as they play Double Compare.

## Counting and Counting On

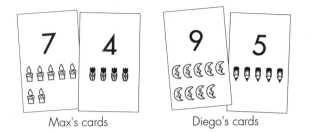

Max's cards          Diego's cards

Max and Diego get right to work finding their totals. Diego counts quietly to himself. He begins at 9, and then counts "10, 11, 12, 13, 14." With each number he says, Diego uses his right index finger to bend back one of the fingers on his left hand. When he has bent back all the fingers on his left hand, he stops counting and announces that he has 14.

Meanwhile, Max is still counting. He began by looking first at the 7 card and counting from 1 to 7. Then, he turned to the 4 card and began counting "8, 9...." When Diego announced his total of 14, Max lost his place. He begins counting again. He counts to 7, and then he counts the pictures on the 4 card, saying "8" as he points to the first picture, "9" as he points to the second, and so on, until he reaches 11. The boys agree that Diego has the greater total.

Like Shavonne, Max puts the two quantities together and counts them all, starting at 1. Diego can begin with one quantity and count on by starting at the next number. In order to do this, Diego treats 9 as a unit; that is, he can think of it as 9 without breaking it down into ones again. Then he counts on—"10, 11, 12, 13, 14"—while keeping track with his fingers of how many he needs to add (1, 2, 3, 4, 5). The teacher feels that the game is at an appropriate level of challenge for Diego and Max. In future sessions, she will observe them to see how their strategies for counting and combining are developing. For example, she will note whether Max continues counting from one each time, and whether they have begun developing strategies for determining particular combinations without counting.

## "Just Knowing" Number Combinations

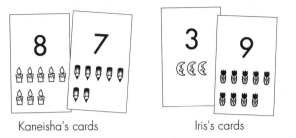

Kaneisha's cards          Iris's cards

As Kaneisha and Iris turn over their cards, Kaneisha immediately announces that she has 15, then looks over at Iris's cards. The teacher reminds Kaneisha to let Iris find her own total. Meanwhile, Iris counts almost inaudibly to herself "10, 11, 12," and then says that her total is 12. Kaneisha says, "Me! I won."

**How did you get your totals?**

**Kaneisha:** Because I know 8 and 7 makes 15. Because it's easy.

**Iris:** I counted in my mind.

Kaneisha's cards          Iris's cards

On the next round, Kaneisha again immediately announces her total, and then waits impatiently while Iris slowly counts on from 9.

When the teacher again asks how the girls found their solutions, Kaneisha is still unable to explain. She seems either to have memorized some number combinations or to have developed strategies for finding solutions to number combinations quickly. As Kaneisha is eager to play at a faster pace than she can with Iris, the teacher decides to ask her to play with Nathan, who is also finding number combinations quickly. To provide further challenge, she may ask Kaneisha and Nathan to turn over three cards on a round.

## Reasoning About Number Combinations

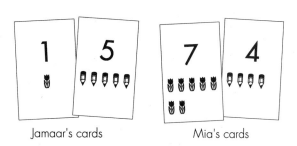

Jamaar's cards        Mia's cards

**Jamaar:** One and 5 is 6. I have 6.

**Mia:** 8, 9... *[after a short pause]* Me! Because you have 6 and I have more. Let's do it again.

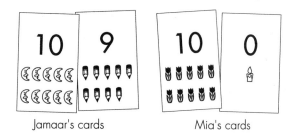

Jamaar's cards        Mia's cards

**Jamaar:** Me! 9 is bigger than 0. You know because it's just your eyes that tell you.

Mia and Jamaar are reasoning about the number pairs without necessarily needing to add them up. Although the teacher has observed in previous sessions that Jamaar and Mia are skilled at counting, comparing, and combining numbers, she feels that this game is deepening their understanding of numbers and number relationships as they explore ways to reason about numbers.

## Choosing a Card to Win a Round

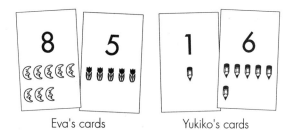

Eva's cards        Yukiko's cards

Eva and Yukiko count together slowly, starting with Eva's cards. Eva touches the pictures as she counts.

Yukiko tells her she has counted a picture twice, and they begin again. After several trials, they complete the count with a total of 12.

**Yukiko:** Six and 1 is 7. But it's my turn to win this time... *[She places her cards side by side so that the 1 is to the left of the 6.]* It's 16. I win!

They both laugh, knowing that the pictures can be combined to find the total number, but the digits cannot be combined in that way.

Next, Yukiko removes her 1 card and places it facedown in her discard pile. She pulls out a 3 card from her pile and then returns it, saying "Three, too small." Then, she pulls out a 2 card from her pile, hesitates, and puts it back, saying "I need something big." Finally, she pulls out a 9, and puts it faceup with the 6.

**Yukiko:** There. It's bigger than your 12. I won.

**Eva:** OK. My turn to win next.

Eva and Yukiko have invented a version of Double Compare in which players win alternate rounds. If the player whose turn it is to win has a losing hand, she can choose a replacement card. Although Eva and Yukiko appear to need more practice counting and combining (they struggled to combine 8 and 5, and arrived at an incorrect total), the teacher decides not to intervene at this point. Although they are not always finding the total of their combinations correctly, they are nonetheless gaining practice counting, comparing, and combining. They are concentrating fully on their work and, as they find winning combinations, they are reasoning about the relative size of numbers and number combinations. The teacher compliments them on developing a collaborative version of the game.

# Breaking Numbers into Two Parts

## What Happens

In On and Off, students toss a set of counters over a piece of paper and record how many land on and off the paper. In Counters in a Cup, one student secretly hides some of a set of counters in a cup, and a partner uses the remaining counters to determine how many have been hidden. These two games and Double Compare are offered during Choice Time. At the end of Session 5, students share the number combinations they got when playing On and Off. Their work focuses on:

- finding combinations of 5 and 7
- using objects to model number combinations
- exploring relationships among combinations of a number
- finding the total of two quantities up to 10
- finding the larger of two quantities up to 20
- using numbers to record how many

## Materials

- Number Cards (1 deck per pair)
- Paper cups (1 per pair)
- Counters such as buttons, bread tabs (at least 7 per student)
- Large display versions of game grids for On and Off, Counters in a Cup
- Student Sheet 7 (1 per student, for homework)
- Student Sheet 8 (1 per student and a few extras for class, plus 1 per student, homework)
- Student Sheet 9 (1 per student, for homework)
- Student Sheet 10 (1 per student and a few extras for class, plus 1 per student, homework)
- Unlined paper for game mats (1 sheet per student)

## Activity

Most of Sessions 4 and 5 will be Choice Time, but first you need to introduce two new games: On and Off, and Counters in a Cup.

In the meeting area, post one copy of your chart version of the On and Off Game Grid (keep the second copy for follow-up work at the end of Session 5). Also have ready your chart version of the Counters in a Cup Game Grid. You'll need a sheet of unlined paper, a paper cup, and about 10 counters. Take seven counters for the first game and leave the rest set aside in the cup.

**On and Off** Place the sheet of paper down as a game mat. Put seven counters on the sheet, and ask how many counters there are. Then, gather them up in one hand.

**I have seven counters. I'm going to toss them over this sheet of paper. What do you think might happen? Let's see how many land on the paper and how many land off the paper.**

### Introducing Two New Games

Toss the counters over the paper in such a way that some will land on, some off the sheet. Ask how many landed *off* the paper. Use the large chart to demonstrate filling in the On and Off Game Grid. Write 7 in the *Total Number* blank, and in the top row of the grid, write how many landed *on* and *off* the paper.

Repeat the activity once or twice, until you think students understand the steps. Each time, write the number landing on and off the paper.

Explain that students will have the game On and Off as one of their choices during Choice Time, with their own game grids to fill in.

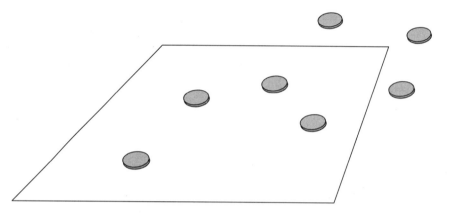

**Counters in a Cup** To introduce the second game, place five counters and an upside-down paper cup in front of you. Ask how many counters you have. Then, explain that you are going to hide some under the cup.

**I have five counters and an empty cup.** *[Hold up the cup so students can see that it is empty.]* **I'm going to hide a secret number of counters under this cup.**

Gather up the counters, and place two of them under the cup in such a way that students cannot see what you are doing. Then, place the remaining counters in front of the cup so students can see them. Motion for students to keep their answers to themselves for now.

**Think silently to yourself about how many counters I hid. When you think you know how many I hid, you can raise your hand, but please don't say anything. We want to give everyone a chance to think about how many counters are under the cup.**

Most students will probably be able to work mentally, but you might make additional counters available for student use if needed.

When most students are ready, ask for volunteers to tell how many they think are hidden.

**How many counters did I hide? How do you know?**

**Tony thinks there are two counters hidden, because he counted the three counters and then counted up two more: 4, 5. Did anyone find their answer in a different way? Does anyone have a different idea about how many counters are hidden?**

Encourage students to explain how they found the answer. Possible strategies include counting the number of counters visible and then counting on to 5; showing 2 fingers on one hand and finding the number needed on the other hand to make up 5; and reasoning that 5 is 2 more than 3.

Be sure to ask for other ideas about how many counters are hidden, even after someone suggests the correct solution. That way, you help students think for themselves about whether they have found the correct solution.

Use the large chart to demonstrate filling in the Counters in a Cup Game Grid. Write 5 in the *Total number* blank, and write the number of counters *inside* and *outside* the cup in the top row of the grid.

Repeat the activity once or twice, until you think students understand the steps. Always start with all the counters out of the cup, where students can see them. Vary the number of counters that you hide. Each time, record the results on your large grid.

Explain that Counters in a Cup is another game for Choice Time that students will be playing in pairs with their own game grids.

## Choice Time

Choices

1. Double Compare ⬜7⬜ ⬜1⬜ ⬜4⬜ ⬜3⬜

2. On and Off

3. Counters in a Cup

Post a list of the choices. Briefly remind students about their options and explain that they will be working on these choices for the rest of today and most of the next session. By about 20 minutes before the end of the two sessions, students should have tried all three choices. If you notice some students working slowly, you might suggest that they complete a certain portion (rather than all) of each game grid.

### Choice 1: Double Compare

**Materials:** Number Cards with wild cards removed (1 deck per pair); cubes or other counters (available)

Students will be familiar with Double Compare from their work in the previous session. Review the rules as needed. Students play in pairs, dealing out a deck of Number Cards evenly. Players simultaneously turn over the top two cards in their piles. The player with the larger total says "Me." If the totals are the same, those cards are ignored and each player turns over two more cards. The game is over when players have turned over all the cards in their piles.

### Choice 2: On and Off

**Materials:** Counters (buttons, bread tabs); Student Sheet 8, On and Off Game Grid (1 per student, plus extras); unlined paper for game mats (1 sheet per student or pair)

Students work alone or in pairs with seven counters and a sheet of paper. They toss the counters and write the number of counters that land *on* and *off* the paper, repeating the activity until they have completely filled a game grid.

If some students are always getting all of their counters on the paper or all off the paper, suggest they adjust the height of their hands before they toss the counters until they get a mix of some on, some off.

**Different Levels of Challenge**  For more challenge, after students have filled a game grid, they can think about the following: The game grid shows different ways to break the number 7 into two parts. Can they think of a way to do this that is not on their game grid?

Students could repeat the activity with a larger number of counters, such as 10.

## Choice 3: Counters in a Cup

**Materials:** Paper cups (1 per pair); counters (buttons, bread tabs); Student Sheet 10, Counters in a Cup Game Grid (1 per pair, plus extras)

Counters in a Cup is a Choice Time option that students must do in pairs. Each pair has a set of counters and a cup. One student hides a secret number of the counters, and the other figures out how many are hidden. They write these numbers on their game grid. Then they switch roles, giving the other student a turn to hide counters. Pairs repeat the activity, taking turns with the hiding, until they have completely filled the game grid. Encourage players to vary the number of counters they hide, but they can occasionally hide the same number of counters more than once.

You will need to decide what number of counters students begin working with. Some teachers decide on a single number of counters (such as 5) that all students work with initially. Other teachers assign different numbers of counters (such as 4, 5, and 6) to different pairs.

**Different Levels of Challenge**  As you observe during this activity, you can adjust the level of challenge for particular pairs by suggesting that they use a larger or smaller number of counters. Counters in a Cup can be challenging for some students, as it involves keeping track of and coordinating three amounts: the total number of counters, the number visible, and the number hidden.

You may find a lot of variation in your class. Some students may need to do this activity several times with 3 or 4 counters before they can work with a set as large as 5. Others may find 5 or 6 too easy, and may need to work with 8 or 9 counters. Once you think students are working at an appropriate level of challenge, be sure they have a chance to do the activity several times with the same total number.

When you see students having difficulty, you might work with them for a couple of rounds. Place all the counters in front of the cup, and ask the students to count them. Then, secretly hide just one counter. Count the visible counters with the students, and remind them of the total. Ask students to try to visualize what's inside the cup:

**There are three counters out here. Can you try to see in your mind what might be inside the cup, so that there are four in all?**

If students are not certain how many are hidden, count the hidden counter as you tap the cup once. Work through the activity with the students again, hiding two counters this time.

Another approach that some students find helpful is modeling the problem with their fingers. They hold up fingers on one hand to show the number of counters visible, and then they find how many fingers they need to raise on the other hand to reach the total.

## Observing the Students

### Double Compare

- Are students reading the numerals or counting the pictured objects on the cards? Do they count accurately?

- How do they combine the quantities on their two cards? Do they count from 1? count on from one of the numbers? "just know" the sum?

- What strategies do students have for deciding which total is greater?

### On and Off

- How do students count the counters on and off the paper? Can they determine the numbers visually?

- Do students count the counters both on and off the paper, or do they use one of the numbers to find the other? ("There are 4 on the paper, and 4 and 3 is 7, so there must be 3 off the paper.") Do they notice if the numbers they have recorded do not sum to 7, for example, if they have counted incorrectly, or if one of the counters is misplaced?

- Do students recognize when the same combinations come up more than once?

- Do students understand how to use the game grid? Do they know how to use 0 to show when counters are all on or all off the paper?

- Are students recording the numbers in the appropriate column? To check students' understanding of the relationship between the game grid and the way the counters landed, you might choose one of the number pairs on their grid and ask students to model it with counters on and off the paper.

**Counters in a Cup**

■ Do students have strategies for finding the number of counters hidden? Do they seem to guess at how many are hidden? If students are having difficulty, refer to the suggestions for Different Levels of Challenge (p. 63).

■ What strategies do students use for finding the number of hidden counters? Do they count by 1's all the counters they see, and then continue counting on to the total number of counters? Do they start with the number of counters visible and count on? Do they use knowledge of number combinations? Do they use relationships among combinations? ("Last time, there were 3 outside of the cup and 2 hidden. This time, there are 2 outside so there must be 3 hidden.")

■ How do students go about varying the number of counters they hide? Do they refer to the game grid to help them decide? Do they appear to have a random or a systematic approach to varying the number of counters hidden? Do they recognize when the same number of counters has been hidden more than once? Do they think of hiding all or none of the counters? (If students do not think of hiding all or none of the counters, do not suggest this yourself. They enjoy discovering this "trick" on their own or learning it from their classmates.)

■ Do students understand how to use the game grid? Do they know how to use 0 to show when all the counters are hidden or visible?

**Activity**

## Sharing Combinations of Seven

About 20 minutes before the end of Session 5, gather students to share the combinations of 7 they came up with when playing On and Off. Post your second large display chart of the On and Off Game Grid. Have a sheet of unlined paper and seven counters available, and ask students to bring their own copies of the game grids they filled in while playing On and Off.

**As I was watching you play On and Off, I noticed lots of different ways the counters can land on and off the paper. Who would like to share one of the ways they found?**

**Claire got 6 on and 1 off. Did anyone else get that? Raise your hands. Did anyone get something different?**

Record each combination of 7 on your large game grid. For two or three of the combinations, ask a student to model the number pair by placing counters on and off the paper. That way, you help students develop visual images of combinations of 7. See the **Dialogue Box,** Combinations of Seven (p. 67), for an example of this sharing in one class.

As the list of number pairs you are recording grows, remind students that you are looking for ways that you have not already recorded. Even if you (or some students) notice that all possible ways the seven counters could land have been recorded, ask once more for other ways to add to the list. That way, you encourage students to look carefully at their own lists. When no one can suggest any other ways that the counters landed, you might ask students if they think could find a way that is not recorded on the class list. In one class, when the teacher asked for this, a student suggested recording a subtraction expression. The teacher encouraged students to think about whether subtraction is an appropriate way to show the On and Off situation. See the **Dialogue Box,** Combinations of Seven (p. 67).

# Sessions 4 and 5 Follow-Up

 **Homework**

**On and Off** After Session 4, students play On and Off with someone at home. Send home a copy of Student Sheet 7, On and Off, and Student Sheet 8, On and Off Game Grid. They will need something they can use as counters, such as buttons, bread tabs, or pennies. You might suggest a total number for students to use, or give them a list of numbers to choose from (such as 6, 8, 9, 10, 11, or 12).

**Counters in a Cup** After Session 5, students teach someone at home how to play Counters in a Cup. Send home a copy of Student Sheet 9, Counters in a Cup, and Student Sheet 10, Counters in a Cup Game Grid. Again, they will need counters, and they will need a paper cup or other small opaque container. Either suggest a particular number of counters for students to use, or offer a list of numbers to choose from (such as 4, 6, 7, 8, 9, or 10).

**Extension**

**On and Off: Finding All the Ways** Students think further about whether the class list contains all possible ways for seven counters to land on and off the paper. You may want to leave your list posted for another day or two and suggest that students who think they have found other ways add them to the list. If some students are exploring this, do not suggest that they organize the class list of number pairs in a particular way, such as from smallest to largest number on the paper. Students need opportunities to find their own ways to organize their work, and they need a chance to begin seeking patterns on their own.

## Combinations of Seven

Students have gathered to share the combinations of 7 they made when playing On and Off. The teacher records each combination on a large copy of the game grid.

**Jacinta:** I have a 4 and 3.

**Four on? or off?**

**Jacinta:** Four on and 3 off.

**Who else got 4 on and 3 off? Raise your hands. Lots did! OK, another one?**

**Diego:** Two on and 5 off.

**Who else had 2 on, 5 off? OK, another one?**

**Brady:** Four on and 3 off.

**Yukiko:** That's already on the list.

**Do you have one we haven't listed yet?**

**Brady:** Seven on.

**Seven on and how many off?**

**Brady:** Zero.

**Hands for 7 on and 0 off? Another?**

**Michelle:** Two on and 6 off.

**Interesting. Up here we have 2 on and 5 off, and you're saying 2 on and 6 off. Can we have both?**

**Michelle:** Oh, that's 8.

**Which?**

**Michelle:** Two and 5. No, 6.

**Everyone figure it out. What's 2 and 6?**

*Some students check by counting on their fingers, others use counters, and others work mentally. Although students are eager to call out the answer, the teacher insists they wait until most of the class has arrived at a solution. Students confirm that 2 and 6 is 8.*

**OK, Michelle, do you have another one?**

**Michelle:** Three on and 4 off.

**Tuan:** We have that. At the top.

**Three on and 4 off. Do we have that?**

**Tuan:** Oh, we have it with 4 first.

**Is that the same or different?**

**Michelle:** It's sort of like it, but it's 4 and 3, and I have 3 and 4.

*Students suggest two more combinations. Then, no one can come up with another way the counters landed.*

**So, you've told me all the ways on your game grids. Are there other ways the counters could land, even if they're not on your sheets?**

**Tamika:** Eight minus 1.

**We're talking about ways to break 7 into two parts. Some counters on the paper, and some off. How could we show that with 8 minus 1?**

**Tamika:** Um, 8 on, and 1 off?

**Let's try it. Who would like to show us?**

Diego volunteers. He models 8 on and 1 off with counters, and notes with surprise that there are 9 counters in all, not 7. Tamika says that it would be 7 if you take away 1 from 8, but that you need to find a way to do it with 7 in all. While a few students seem to recognize why 8 minus 1 is not a solution—they are beginning to see that game is not a subtraction situation—the teacher is not sure that everyone is following. She decides that for now, she will not pursue this any further, but will leave the chart up for students to think about.

| ON | OFF |
|----|-----|
| 4 | 3 |
| 2 | 5 |
| 7 | 0 |
| ~~2~~ | ~~6~~ |
| 3 | 4 |
| 6 | 1 |
| 0 | 7 |
| ~~8~~ | ~~1~~ |

# Towers of 10 and Number Choices

## Materials

- Interlocking cubes (class set, grouped by color)
- Paper cups (6–8 for the class)
- Dot cubes (12–16 for the class)
- Counters (10–15 per pair)
- Overhead projector
- Dot card transparencies
- Student Sheets 8 and 10 (continuing supply available for Choice Time)
- Game Record Sheet (1 per student, homework, optional)

## What Happens

In the game Towers of 10, students use cubes of two colors to build towers 10 cubes high, then record the number of each color in each tower. Towers of 10 is added to the other games for Choice Time each day. Session 7 starts with more Quick Images, and at the end of Session 8, students share their combinations of 10 from playing Towers of 10. Their work focuses on:

- finding combinations of numbers up to about 10
- using objects to model number combinations
- exploring relationships among combinations of a number
- recording solutions with pictures, numbers, and words

## Activity

### Introducing Towers of 10

Towers of 10 will be one of the choices during Choice Time in Sessions 6–8. To introduce the game, bring two dot cubes and about 60 interlocking cubes in two colors to the meeting area. Ask for a student volunteer to play a demonstration game with you. If you have a large class or think some students will have difficulty seeing the towers you build, keep track of the progress of the game by drawing towers on chart paper or the board as you build them. Have counters available for students to use to figure out the totals in the demonstration game.

**We're going to learn a new game called Towers of 10. You and your partner will be building three towers together. Each tower is ten cubes high. To start with, each player chooses one color of cubes to use.**

Ask the student you are playing with to choose one of the two colors, and explain that you will use the cubes of the other color.

**Jacinta chose blue, so I'll use the yellow cubes.**

Then demonstrate the first turn by rolling the dot cubes.

**We take turns rolling the two dot cubes. What numbers did I roll?...
I rolled 5 and a 4. So, how many cubes do I take? How do you know?**

Make a stack of nine yellow cubes. If you are showing the progress of the
game on a piece of chart paper, draw a tower of nine yellow cubes.

**So, I used my cubes to make a tower nine cubes high. How many more
cubes do we need to make a tower of ten?**

Encourage students to count on their fingers or try to visualize the situation
if they find it helpful, but keep in mind that some students may find this
question too difficult. Students will gain more experience with combinations
of ten when they play the game.

**Note:** If you roll a 5 and a 6 or two 6's on your first turn, put the cubes
together to make a tower, and then ask students how many cubes you
should snap off so that you have only ten in the tower.

Next, your student partner rolls the dot cubes.

**Jacinta rolled a 3 and a 2. How many blue cubes does she take?**

Show the student how to put one blue cube on top of the tower of nine
yellow cubes to make a complete tower of ten. Then start a new tower
with the remaining four cubes. If you are showing the progress of the
game on a piece of chart paper, update your drawing by adding a blue
cube to the tower of nine and drawing a tower of four blue cubes.

**Now we've finished one of our three towers of ten. We've also started
a tower of four. How many more do we need to finish that tower?**

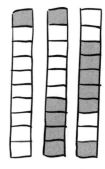

Again, don't spend too much time on this question, as it may be difficult
for some students.

Continue playing until you have completed three towers, each ten cubes
high. You don't need to roll an exact number to finish the game; extra
cubes are set aside.

The last step is to find the number of cubes of each color in each tower.
Choose one tower to start with, and count aloud as students count the
number of each color with you. Repeat this count with each tower.

**How many yellow cubes in this tower?... Yes, there are 4 yellow cubes:
2 in the middle, and 2 at the top. How many blue cubes?... Yes, there are
6 blue cubes: 1 at the bottom, and then this group of 5. So, this tower of
10 cubes has 4 yellow cubes and 6 blue ones.**

## Choice Time

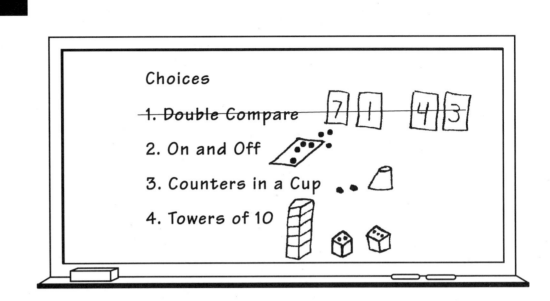

Choices

1. Double Compare ~~7 1    4 3~~
2. On and Off
3. Counters in a Cup
4. Towers of 10

Add Towers of 10 to your posted list of choices and cross off Double Compare. Tell students that by the end of this Choice Time (which runs for three sessions), they should have done all three choices that are still on the list, either in this Choice Time or the previous one.

For a review of the descriptions of Choices 2 and 3, On and Off and Counters in a Cup, see pp. 62–63. You will need to decide what numbers students work with for these two games, based on your observations of student work during Sessions 4 and 5. You might ask all students to use the same number of counters (say, 8), or you might assign different numbers to particular pairs, to provide more or less challenge. Some teachers offer a set of numbers from which students can choose for each activity:

For On and Off:  6, 8, 9, 10, 11, or 12

For Counters in a Cup:  4, 6, 7, 8, 9, or 10

### Choice 4: Towers of 10

**Materials:** Interlocking cubes (about 60, in two colors, per pair); dot cubes (2 per pair); lined or unlined paper for recording

Students play in pairs, building three towers of ten cubes each. They each choose one color of cubes, then take turns rolling the dot cubes and snapping together the number of cubes rolled, in their chosen colors, to form towers. They work together on the same towers, not starting a new one until one tower of ten is finished. Students continue taking turns until they have three complete towers.

Then, both students record how many cubes of each color are in each tower. They may record with pictures, numbers, words, or a combination. Students who are not able to write the color names can use crayons or markers to show the colors. Each pair plays the game twice.

**Different Levels of Challenge**  If some students seem to have difficulty working with ten, suggest they build shorter towers (for example, towers six cubes high).

Students ready for more challenge can play to make five towers, or they can make towers 11 or 12 cubes high. They might also think about the following question: The recording sheet shows different ways to break the number ten into two parts. Can they think of a way to do this that they have not already written down?

## Observing the Students

See pp. 64–65 for guidelines on observing students' work during Choice 2, On and Off, and Choice 3, Counters in a Cup.

### Towers

- Do students find the total of the two numbers they roll, or do they just take cubes for the number on each dot cube? If they find the total, do they count each dot? Do they count on from one of the numbers rolled? Do they use knowledge of number combinations? Can they quickly add on 1 or 2 to another number?

- How do students count the number of cubes in the towers? Do they count each cube separately? Do they count on from one of the groups in the tower? ("There are 4 blue cubes and the rest are red, so that's 5, 6, 7 in this tower so far.") Do they use knowledge of number combinations? ("There are 4 blue cubes and 3 reds, so the tower has 7.")

- Can students determine how many more cubes they need to complete a tower? What strategies do they use?

- How do students find the number of each color in their towers? Do they count all the cubes? Do they use knowledge of combinations of 10? ("There are 4 green cubes, so there must be 6 yellows.") Do they notice if they have made a "completed" tower that does not have 10 cubes? Do they notice that all completed towers are the same height?

- How do students record their work? If some students just record the number of cubes in each color (for example, "4 and 6"), remind them they need to show what color each number stands for. How do students record none (0) of one color?

## Quick Images

At the start of Session 7, do Quick Images with the class two or three times. Show any dot card transparencies you have not previously used, or use transparencies or enlargements of the dot pictures students made in Investigation 1, or dot pictures you make up yourself. See page 6 for a review of the Quick Images activity.

## Sharing Combinations of Ten

Ask students to wrap up their work on Choice Time about 20 minutes before the end of Session 8. Call the class together to share the combinations they got when playing Towers of 10. Have interlocking cubes of two colors at the meeting area, and ask students to bring their recording papers.

Ask for volunteers to tell the number of each color they used in one tower. Students will probably include the colors when they describe their towers, saying "3 blue and 7 red" rather than "3 and 7." Help them focus on the numbers rather than the colors:

**Chanthou and Brady made a tower with 3 red cubes and 7 blue ones. Did anyone else make a tower with 3 of one color and 7 of another? You might have used other colors, but did you get 3 of one and 7 of another?**

**Who made a tower with different numbers of each color?**

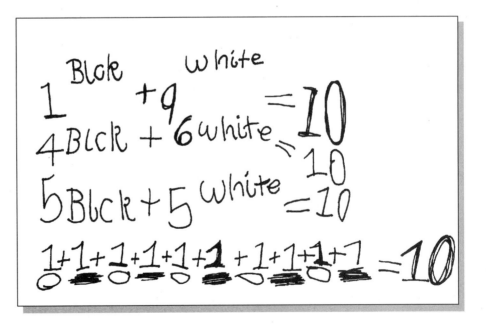

On chart paper or on the board, record combinations in the order in which students suggest them. Each time someone suggests a combination, ask the class to check if it is already listed. Students may ask if the order of the numbers matters, that is, whether "4 and 6" is the same as "6 and 4." Encourage the class to share their ideas about this. Remind them that in this situation, you're mostly interested in the two numbers—it doesn't matter if it's 4 yellow and 6 green, 4 green and 6 yellow, or 4 green and 6 blue.

Ask students if they can think of a game they played recently in which the order of the numbers did matter. For example, in On and Off, the first number they recorded showed how many *on* the paper, and the second number showed how many *off*. In Counters in a Cup, the first number they recorded showed how many *in* the cup and the second number showed how many *out of* the cup. By drawing attention to situations in which order matters and situations in which it doesn't, you help students see how we use numbers and number combinations to model real events.

Continue accepting and recording combinations of ten until students can no longer suggest any new ones, or until students begin to seem less attentive. (Some students may find it hard to stay focused once all the combinations of ten on their own recording sheets are listed.)

$$3 + 7$$
$$6 \times 4$$
$$9 + 1$$

E Black and 7 red
5 red and 5 Black
4 red and 6 Black

10 Black and o red
10 red and o Black
5 red and 5 Black

# Sessions 6, 7, and 8 Follow-Up

 **Homework**

**Math Games** Students have taken home five games during this unit: Compare Dots, Copying Counters, Double Compare, On and Off, and Counters in a Cup. They will benefit by continuing to play these games for homework and as choices throughout the unit. See the **Teacher Note**, Games: The Importance of Playing More Than Once (p. 170), for more about the value of playing the games in the *Investigations* curriculum multiple times.

You may want to send home the Game Record Sheet (p. 225) and either assign a particular game or list games from which students can choose. This sheet helps students reflect on the games they are playing at home, and helps them understand that practice at home is an important part of the mathematics work they are doing.

If you send home the Game Record Sheet, talk briefly about telling the story of what happened in a game. For example, one response after playing On and Off might be "We played with paper cups. We used 10 in all. We got 4 on, 6 off the most. It came up 7 times!" Students fill out the sheet with family help and return it the next day.

**Extension**

**All the Combinations of 10** Students think about whether the class list you made at the end of Choice Time contains all possible combinations of 10. You might leave this list posted for another day or two, and suggest that students who think they have found other combinations of 10 add them to the list.

# Three Towers

## What Happens

**Materials**

- Interlocking cubes (50 per student, 25 each of two colors)
- Lined or unlined paper for recording
- Counters (available)

As an assessment, individual students use cubes of two colors to make three different 10-cube towers. They record the number combinations represented by their towers. For the rest of the session, students solve a How Many of Each? problem about 15 things. Their work focuses on:

- finding combinations of 10
- using objects to model number combinations
- recording solutions with pictures, numbers, and words

**Activity**

**Assessment**

**Three Towers**

Note: Because each student needs 50 cubes for the assessment activity, you may need to have part of the class start on the activity Fifteen in All while the rest work on the Three Towers assessment. As some students finish with their cubes, break them apart and distribute them to students who haven't yet done Three Towers.

Introduce the assessment activity and, as needed, explain the cube supply situation. Then quickly introduce the next activity, Fifteen in All, for students to work on while they are waiting for cubes. Distribute the cubes, two colors to each student, as you begin your introduction.

**During Choice Time over the last few days, you've played the game Towers of 10. I've seen you make lots of different towers of ten with your partners. Today, each one of you is going to make three towers on your own.**

**Use the two colors of cubes you have in front of you. Make three towers, each ten cubes high. Use a *different* number of each color in all three towers.**

**It's like this: If I have yellow and blue cubes, I could put 3 yellow cubes in one tower. But for the other two towers, I couldn't use a total of 3 yellows. I'd have to use some other number of yellow cubes.**

Explain that just as when they played the game Towers of 10, they record the number of each color they use in their towers when they have finished.

## Observing the Students

During this assessment, observe how students are thinking about combinations of ten, and notice the ways that they count and combine sets. Ask students to explain their thinking if you are unsure of their strategies. See the **Teacher Note,** Assessment: Three Towers (p. 79), for more suggestions for assessing students' work on this activity.

■ How do students approach the task? confidently? cautiously? In need of further support or direction from you?

■ What strategies do students have for making their towers? Do they put one cube at a time on their towers, perhaps counting the whole tower with each new cube added, until they have ten in all? Do they use their knowledge of number combinations? Do they use relationships among combinations? ("I did 5 red and 5 green, so next I'll do one less red— 4 red and 6 green.")

■ What strategies do students have for counting and keeping track of the numbers of cubes in their towers? Do they count each cube individually? Do they count groups of cubes of different colors? ("There are 2 blue cubes, then 3 more blues at the bottom of the tower, that makes 5.")

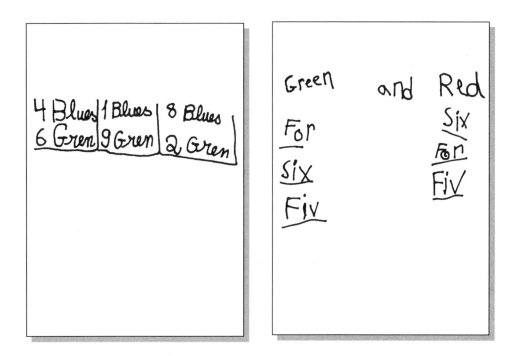

■ How do students record the number of each color in their towers? Can they use numbers to show the number of cubes accurately?

■ Do students have strategies for determining if their solutions are correct?

If you have plenty of cubes, students who finish early can find additional towers of ten that use different numbers of each color.

## Fifteen in All

For the alternative activity, which everyone does after finishing Three Towers, students work on another How Many of Each? problem. The suggested total is 15, although you may choose another number for certain students or for the whole class. It is fine for students to repeat a total they have already used in How Many of Each? problems. They benefit from repeated practice with counting and number combinations, and they may notice new relationships among solutions.

Consider basing the problem on one of the stories students made up in the activity Problems about Ten (Session 2). Or, make up your own problem. For example:

**I have 15 chickens and pigs in a barn. What could I have? How many of them are chickens? How many of them are pigs?**

On the board or on a piece of chart paper, write "15 chickens and pigs."

❖ **Tip for the Linguistically Diverse Classroom** Sketch simple pictures for whatever two items you choose for this problem.

Students may work alone, in pairs, or in small groups. Make available counters, cubes, or other materials to help solve the problem. When they have found a solution, they record it on unlined paper using pictures, numbers, words, or a combination of these. Their solution should include the number of each kind of thing and the total number of things, either in numbers or words.

If time permits, at the end of Session 9, take a few minutes for students to share how they recorded their work.

**I saw that you found different ways to record your work on this problem. Some of you used pictures, some of you used numbers, and some of you used words. Does anyone want to share how you recorded?**

Call on students who have recorded in different ways. For example, you might call on a student who used numbers, one who used equations, and one who used pictures and numbers. As students share, do not discourage them from telling their solutions, but keep the focus on how they recorded their work. Students might keep working to find more solutions later in the day, the next day, or at home.

# Assessment: Three Towers

As you observe students working on the assessment activity Three Towers, jot down notes about how each student is approaching the task, using the questions on pp. 76–77 as guidelines. Students' written work will give you a sense of how they are recording combinations but not how they are approaching the task. It is particularly important to observe each student working on this assessment to learn about how they are thinking about combinations of 10.

You may find that students fall into three general groups:

Many of your students are likely to be doing competent and appropriate work. They have strategies for solving the problem and can record their work accurately. Although they may know some combinations of 10 and may even recognize some relationships among combinations, they often use strategies that involve counting by 1's to determine the total number of cubes and the number of each color. For example, to create a tower, they might connect a few cubes of one color, count them, and then continue adding more cubes of another color and counting on by 1's until they have a total of 10. Or they might work mentally, choosing a number to represent one color, and then using their fingers to count on from that number in order to determine how many they need of the other color.

Another group of students will more consistently use strategies that involve combinations of 10. You might see them take 4 cubes of one color and 6 cubes of another, without counting the total, because they know that 6 and 4 is 10. Some might solve the problem without building any towers, explaining that they know that 5 and 5, 3 and 7, and 2 and 8 are all ways to make 10. Some of these students may even be comfortable breaking 10 into more than two parts, for example, they might explain that they know 2, 2, and 6 is another way to make 10, because you can break the 4 into 2 and 2. To challenge these students further as they work on this unit, you

might suggest that for number combination activities (such as How Many of Each? problems), they use larger numbers, or they use more than two addends.

In a third group will be those students about whom you have some concerns. They may be uncertain of how to approach the problem, or they may have difficulty keeping track of and coordinating the total and the two parts. For example, they may not readily see why 3 yellows and 6 greens *and* 3 yellows and 7 greens could not both be towers of 10. These students will nearly always count by 1's, and their counting may not always be accurate. Some of these students might benefit from further experiences with combinations of numbers under 10 or in the lower teens. As they move on to new activities, problems, and games in Investigations 3 and 4, provide opportunities for them to continue playing the games Counters in a Cup and On and Off, both for homework and outside of math time.

# INVESTIGATION 3

# Counting

## What Happens

**Sessions 1 and 2: The 100 Chart** Students explore the Hundred Number Wall Chart. They play Missing Numbers and figure out which numbers you have removed from the chart. During Choice Time, students again play Missing Numbers and also make counting strips by writing the sequence of numbers, starting at 1, on long strips of paper.

**Sessions 3 and 4: Which Holds More?** Students continue Choice Time after the introduction of two new choices: Which Holds More? and Exploring Calculators. In Which Holds More? students estimate which of two containers holds more cubes, and check their estimates by filling them and counting the cubes.

**Sessions 5, 6, and 7: More Counting and Comparing** Students gather at the Hundred Number Wall Chart to play a group game of Missing Numbers. Then they learn to play Ten Turns, rolling number cubes and accumulating pennies or counters. For most of these sessions, students continue Choice Time with Missing Numbers, Which Holds More? Exploring Calculators, and Ten Turns. At the end of Session 7, the class discusses common errors in sequences of written numerals on their counting strips.

**Session 8: Clapping Patterns** In Clapping Patterns, an activity that is repeated throughout grade 1, students act out repeating patterns with a set of physical actions. They also show the pattern in one or more other ways: with drawings, with numbers, or with objects like cubes or pattern blocks.

**Session 9 (Excursion): Counting Stories** Students listen to a counting story. They estimate and count some of the quantities in the book, and find the corresponding numbers on the Hundred Number Wall Chart. Each student then illustrates one page for a class book of a similar counting adventure.

**Routines** Refer to the section About Classroom Routines (pp. 172–179) for suggestions on integrating into the school day regular practice of mathematical skills in counting, exploring data, and understanding time and changes.

## Mathematical Emphasis

- Reading, writing, and sequencing numbers to 100
- Counting quantities up to about 40
- Finding the total of two quantities, one that's just a few and another up to about 40
- Exploring patterns in the number sequence
- Using numerals to record how many, for quantities up to about 40
- Finding the larger of two quantities up to about 40
- Exploring calculators as a mathematical tool

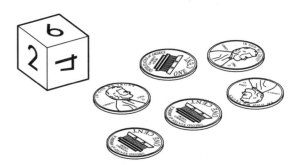

## What to Plan Ahead of Time

### Materials

- Hundred Number Wall Chart with transparent pockets, numeral cards, and colored plastic chart markers (Sessions 1–8)

- Plastic Hundred Number Boards with removable number tiles: 2 or 3 per class (Sessions 1–7)

- Adding machine tape: 1 roll (Sessions 1–4)

- Clear tape (Sessions 1–4)

- Interlocking cubes: class set of 1000 (Sessions 3–8)

- Calculators: at least 6–8 for the class (Sessions 3–7)

- Number cubes: 1–2 per pair (Sessions 5–7)

- Plastic pennies: 50–60 per pair (Sessions 5–7, optional)

- Pattern blocks: 1 bucket (Session 8)

- *Out for the Count: A Counting Adventure* by Kathryn Cave (Simon and Schuster, 1992), or similar counting book (Session 9)

- Drawing paper, crayons or markers (Session 9)

- Unlined paper (available)

- Counters such as buttons, bread tabs, or pennies: at least 30 per student (available for use as needed)

- Chart paper or newsprint (18 by 24 inches): 15–20 sheets (available for use as needed)

### Other Preparation

- For Sessions 1–4, cut the adding machine tape into strips about 2 feet long, making two strips per student.

- Collect about a dozen empty, nonbreakable containers (such as plastic cups, large and small margarine tubs, soup cans, coffee cans) that hold from 10 to 40 interlocking cubes. Label each container with a letter. Determine how many cubes each container holds and keep a list for your reference. (Sessions 3–7)

- Before Session 3, draw a large version of Student Sheet 12, Which Holds More? on chart paper or newsprint.

- Duplicate the following student sheets and teaching resources located at the end of this unit. If you have Student Activity Booklets, copy only items marked with an asterisk.

*For Sessions 1 and 2*

Student Sheet 11, Numbers at Home (p. 202): 1 per student, homework

*For Sessions 3 and 4*

Student Sheet 12, Which Holds More? (p. 203): 1 per student

100 Chart* (p. 206): 2–3 per class, optional

Game Record Sheet* (p. 225): 1 per student, homework, optional

*For Sessions 5, 6, and 7*

Student Sheet 13, Ten Turns (p. 204): 1 per student, homework

Student Sheet 14, Ten Turns Game Sheet (p. 205): 1 per pair for class, plus 1 per student, homework

*For Session 8*

Blank 100 Chart* (p. 207): 10–15 for the class

Cube Pattern Strips* (p. 208): 6–8 sheets for the class, each cut apart into four strips

# The 100 Chart

## Materials

- Hundred Number Wall Chart with numeral cards, plastic chart markers
- Plastic Hundred Number Boards with removable number tiles (2 or 3 per class)
- Adding machine tape cut in 2-foot strips (2 per student)
- Clear tape (available)
- Student Sheet 11 (1 per student, homework)

## What Happens

Students explore the Hundred Number Wall Chart. They play Missing Numbers and figure out which numbers you have removed from the chart. During Choice Time, students again play Missing Numbers and also make counting strips by writing the sequence of numbers, starting at 1, on long strips of paper. Their work focuses on:

- reading numbers
- ordering written numbers
- writing numbers sequentially
- locating numbers on the 100 chart
- exploring patterns in the number sequence

## Activity

### Exploring the 100 Chart

**Note:** If you do not have the Hundred Number Wall Chart from the grade 1 materials kit, you can make one from heavy poster board, paper fasteners, and white and colored index cards. Attach 100 paper fasteners evenly across the poster board in 10 rows of 10. Cut 50 white index cards in half and punch a hole at the top of each one. Number the cards 1–100 and hang each card from a paper fastener. Do the same with index cards in another color; these can be used to highlight certain numbers.

Gather students where they can all easily see the Hundred Number Wall Chart, hung for display. As students look at the chart, start with an open-ended question and then follow the students' lead in choosing follow-up questions. For example:

**What can you tell me about this chart?**... Tony says it goes up to 100. **Who can show me where 100 is on the chart? What else can you tell me about the chart?**... OK, it starts at 1. **What else?** Libby says the first row goes up to 10. **How many numbers are in that row?**

Some students may notice patterns in the columns; for example, that all the numbers in the far right column end in a 0. Others may notice patterns in the rows: the leftmost digit is the same for all but the last number in the row, or each row of numbers repeats the pattern _1, _2, _3, _4…. Recognizing and talking about these patterns helps students learn the sequence of number names and numerals in our number system. See the **Dialogue Box,** Exploring the 100 Chart (p. 90), for the observations of one class and ways the teacher followed up their observations.

After a few minutes of general exploration, start asking students to locate specific numbers on the chart:

**Who can find 10? Who can find 20? Can someone point to 50? Can someone point to 15? Who can find 63?**

This will give you a preliminary idea of how familiar your students are with the sequence of written numerals. Do they have a general idea of where to find a number like 39? Do they count by 1's until they reach 39, or do they just search the chart randomly to find it? If students are not yet very familiar with written numbers above 20, focus most of your questions on numbers up to 20.

**Note:** Many teachers are accustomed to using number lines to explore numbers and number patterns. We focus on a 100 chart instead, because the arrangement of numbers in rows of 10 highlights the base 10 structure of our number system. The structure of the 100 chart can also help students attend to number patterns based on tens. However, both the number line and the 100 chart are useful tools.

If you have a number line posted in the classroom, ask students to think about how it differs from the 100 chart. You may want to make both the 100 chart and a number line available for student use when they are finding totals as they will do in the activity Ten Turns (p. 100) and when solving the story problems in Investigation 4. It is important, however, not to substitute the number line for activities that are specifically built around a 100 chart, such as Missing Numbers (p. 84) and Clapping Patterns (p. 106).

---

# Missing Numbers

Missing Numbers is a game that involves locating numbers on the 100 chart. You can play often with the class as a whole, and students will play this game in pairs during Choice Time in the next few sessions.

Introduce the game with the whole group using the Hundred Number Wall Chart. Remove five cards from the chart and put them facedown out of sight. Have the numeral cards for these numbers nearby, along with red plastic markers, and insert both in the slots as students decide what's missing. For the first round, you might choose a pattern of missing numbers such as 11, 13, 15, 17, 19, or 10, 12, 14, 16, 18.

Point to one of the empty slots (not necessarily the first one in numerical sequence).

**Think silently in your head what number you think is missing from this spot. When you think you know what it is, just look at me like you're ready to tell me.**

Pause long enough to give everyone time to think about what number is missing. Then ask:

**Who wants to tell me what number you think is missing?... Why do you think it's 16? Claire says she thinks it's 16 because it comes after 15. Did anyone else do something like that? Who thought about it a different way?**

As students offer their ideas, encourage others to raise their hands if they used a similar strategy. Then, ask if anyone found the number in a different way. See the **Dialogue Box,** Finding Fifteen (p. 91), for strategies that students in one class used.

When no one can suggest another strategy, ask for a volunteer to write the missing number on the board. It's sometimes hard for first graders to remember how to write the teen numbers. This is a good chance to talk about that and, if it comes up, to discuss how it's easy to confuse two numbers with the same digits in different order, such as 16 and 61.

Put the numeral card for 16 in the slot and insert the red marker over it so that 16 stands out from the other numbers. Continue with the other missing numbers. As students identify the number missing in each slot, replace the numeral card with the red chart marker over it. If students get restless, move more quickly through these first numbers.

When all the numbers have been guessed, ask students if they see any pattern in the red and black numbers.

**After 18, which is the next number that should be red? Then which number? What about the numbers before 10?**

Cover the numeral cards with red chart markers as students direct you. You need not finish the whole pattern, but perhaps complete it up to about 30.

For the next round, first take out all the red chart markers. Then remove five different numbers, this time not in any pattern, just five numbers that you think students will be able to figure out. Ask students to tell you which number they think belongs in each blank space and to explain why they think it's that number.

Explain that students will be doing this Missing Numbers activity again in pairs, either at the Wall Chart or with smaller plastic Hundred Number Boards with removable tiles, during Choice Time.

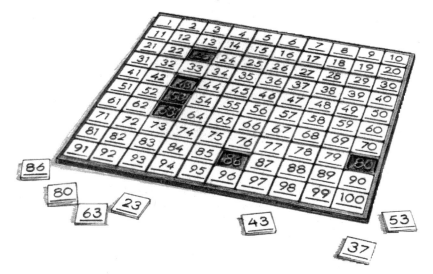

## Introducing Counting Strips

Show students the 2-foot strips of adding machine tape.

**Today and tomorrow, during Choice Time, everyone will spend some time making a counting strip by writing numbers on one of these long strips of paper.**

Ask students how high they think they can count and write numbers.

**Some of us are just learning to write the numbers. Others may already know a lot of numbers. On this strip, you begin with the number 1 and write the numbers, one under the other, as high as you can count. Write each number large enough so that it is easy to read. If you need more paper, you can tape two strips together after one is filled up.**

Tape a 2-foot length of adding machine tape to the board vertically, and demonstrate how to make a counting strip by writing the numbers from 1 to about 12 or 13, one under the other.

It is important to position the strip vertically and write the numbers one under another. When young students try to work horizontally on a strip, their numbers run together and are hard to distinguish.

Suggest that students write their name on the top of their strip.

## Choice Time

Post the Choice Time list for the rest of Sessions 1 and 2. Most students will begin working on their counting strips while pairs (or, if necessary, groups of three) take turns playing Missing Numbers. You will probably want to assign who starts with Missing Numbers and establish an order in which pairs take turns. Counting strips is an easy activity to interrupt and return to; simply show students how to roll up and store their strips when they stop working on them.

## Choice 1: Missing Numbers

**Materials:** Hundred Number Wall Chart with numeral cards in place; plastic Hundred Number Boards with removable number tiles (2–3 available)

One pair can work at the Hundred Number Wall Chart; other pairs use the smaller plastic boards with removable tiles. To start the Missing Numbers game, one student in the pair shuts his or her eyes while the other student removes five number cards or tiles and places them facedown so that the numbers are not showing. The student may remove either random numbers or numbers that make a pattern. Those students using the plastic boards should replace each "missing" number with a blank tile.

The student who removed the numbers points to one of the empty spaces, and the other student says what number belongs there. When the correct number is named, that card or tile is replaced. A pair continues in this way until all the empty spaces are filled.

**Different Levels of Challenge** If some students are finding this activity challenging, ask them to choose only smaller numbers. Encourage them to share their strategies with one another and to think about how they could use the numbers on the chart to help them:

**Are there any clues to help you think about the numbers missing? What about the numbers nearby?**

For more challenge, students could remove larger numbers or they could remove more than five numbers. They might also remove several consecutive numbers and ask that numbers in the middle of the set of empty spaces be identified first.

**Note:** Some teachers have tried to substitute paper 100 charts and use counters to cover the numbers for Missing Numbers. However, this alternative is not very satisfactory as the covering pieces are easily bumped off, revealing the numbers. It's better to work with the wall chart and plastic boards only. With limited materials, you might ask that students take only two turns each at removing numbers, to give more pairs a chance.

## Choice 2: Counting Strips

**Materials:** Adding machine tape in 2-foot strips (1–2 per student); tape

Starting at the top of the strip of adding machine tape, students write their names and then the number sequence as far as they are able. If they fill up one strip and want to continue, they can tape on another length of adding machine tape. Students should completely fill one strip before they add a second strip.

## Observing the Students

As you watch students work during Choice Time, you will be able to get a good sense of how fluent they are in reading numbers, writing numbers, and knowing which number comes next in sequence.

### Missing Numbers

- How do students decide what number is missing? Do they count from 1? Do they count from some other number? Do they use the numbers before and after the missing number?

- Do students use the structure of the 100 chart to help them locate numbers? For example, do they know that a missing number is in the 20's because it's in the row that extends from 21 to 30? Do they know that a missing number must end in 3 because the others in the column do?

- What range of numbers are students comfortable with? numbers under 20? under 40? Do some students have particular difficulty with the teens, but find it easier to locate numbers in the 20's or 30's? Are some students comfortable with numbers anywhere on the chart?

### Counting Strips

- How far in the counting sequence are students able to write numbers? Do they reverse digits in some numbers, for example, writing 31 for thirteen or 04 for forty?

- Are students able to make the transitions between decades? For example, do they know what comes after 19 or 29 or 39?

- Do students recognize any patterns in the counting sequence? Do they use these patterns as they write the numbers?

Take note of common errors that students are making for a later discussion of these. (You will also have the chance to do this when you collect students' counting strips at the end of Session 4.) Keep in mind that errors are to be expected; writing the number sequence correctly is a complex skill that young students will develop over time.

Some students, in particular those who extend their counting strips to numbers over 100, may ask how to say some of the numbers they write. Encourage students to reason this out for themselves. In one class, a student continued working on his counting strip outside of math time and ended up with numbers to 1000. He asked if "ten hundred" was the correct term for the written number 1000. The teacher asked the student why he thought that might be correct, and he explained that the next hundred after 9 hundred should be 10 hundred. The teacher told him that was good thinking and confirmed that the number 1000 is made up of ten hundreds. Once students have explained their own reasoning, they may no longer feel the need to know more. But if they do, do not hesitate to tell them that this number is also called *one thousand*.

Near the end of each session, or when students have completed their work with counting strips, they can roll them up and secure them with a paper clip to make them easier to handle and store.

## Sessions 1 and 2 Follow-Up

**Numbers at Home**  Distribute Student Sheet 11, Numbers at Home. Students are to find a two-digit number on something at home. They write the number and draw a picture of (or explain in writing) where they found it. You might want to brainstorm with the class different places where they might find numbers: near the front door or mailbox, on a food package; on a clothing label; on a postage stamp; on digital displays.

**Homework**

## Exploring the 100 Chart

Students are gathered around the Hundred Number Wall Chart, sharing their observations. The teacher carefully poses questions based on what she hears students say. Many students are eager to respond to her questions and to contribute their ideas. To help them listen to what their classmates are saying and to give more students a chance to participate, the teacher sometimes calls on one student to demonstrate another's suggestion.

**So, what can you tell me about this chart? What is it?**

**Chris:** It has windows in it.

**What do you see in the windows?**

**Diego:** Numbers. It's a number chart.

**Jamaar:** It starts with 1.

**And ends with?**

**Shavonne:** 100.

**Where are the 1 and the 100? Who can show us the 1? Who can show the 100?**

*[Iris and William come to the chart and point out the 1 and 100.]*

**Tuan:** I see something. When you go down [a column] they have the same number on each one. With a different number in front of it.

**Can you come up and show us?**

**Tuan:** *[He points to the rightmost 1 in 11.]* This one has a 1, and this one *[points to the 1 in 21]* has a 1. *[He then points to the 1 in 31 and 41.]* It keeps going.

**Chanthou:** All of them have 1 on this side *[she motions down the first column]*. All 1's, then all 2's in the next one, all 3's, all 4's, all 5's, all 6's, all 7's, all 8's, all 9's, and all 0's.

**Who sees what Chanthou is saying? Does someone want to come up and show us?**

*[Andre comes up and traces down the ones digit of each column.]*

**Yukiko:** If you look carefully you might see... Can I show? *[She comes up to the chart and points to the diagonal made by 11, 22, 33, 44, 55, 66, 77, 88, 99.]*

**Mia:** It goes 1, 1; 2, 2; 3, 3; 4, 4 ...

**And those run on the... what's it called?**

**Jonah:** Diagonal.

**Iris:** At the end *[she points to last number in the top row]* it's 10, and if you keep on going down, you get to 100. If you go the other way *[she motions to indicate across rows]* it takes a lot longer to get to 100. Longer than 20, 30, 40, 50, 60, 70, 80, 90, 100.

**If you're counting like that—10, 20, 30, 40, 50, 60, 70, 80—what number are you counting by?**

**Kaneisha:** Ten.

**When you are counting 1, 2, 3, 4, 5, what are you counting by?**

**Claire:** Ones.

**Garrett:** When you are counting by 10's, like Iris said, you don't really have many numbers, just one row going down, and by 1's there's a lot more. *[Note that some students say "row" for column; as with other new vocabulary, simply model the correct terms as you use them.]*

**Michelle:** I have a new thing. There's 10 rows in all.

**Let's count the rows.**

*[The teacher points to each row from top to bottom, and students count along with her.]*

**Chanthou:** It's like a pattern. I just found out. It keeps on going 10 across. *[She points to each number in the first row, counting]* 1, 2, 3, 4, 5, 6, 7, 8, 9, 10, *[then points to each number in the second row, counting again]* 1, 2, 3, 4, 5, 6, 7, 8, 9, 10.

**Andre:** There's 10 numbers in each row.

**Libby:** And 10 rows, too!

# DIALOGUE BOX

## Finding Fifteen

To introduce the activity Missing Numbers (p. 84), the teacher has removed 11, 13, 15, 17, and 19 from the Hundred Number Wall Chart.

The discussion that follows shows the many different strategies students used to find 15. As students share their strategies, the teacher encourages them to think about whether their own approaches are similar to or different from those their classmates described.

**Who can tell us one of the numbers that's missing?**

Jacinta: 15.

**Can you show us where it would go?** *[Jacinta comes up and points to the appropriate blank.]* **How did you figure it out?**

Jacinta: Because 14, 15.

**Did anyone think about the number before, the way Jacinta did?** *[Several students raise their hands.]* **Who did something different?**

Nathan: I looked at this one *[comes up and points to 5]* and I knew 15 should come right underneath it.

**How did you know that?**

Nathan: Because it goes 5, 15, 25, 35, to 95.

**Anyone else use that pattern?** *[Again, several hands go up.]* **Did anyone think about a different pattern?**

Donte: 5, 10, 15, 20, 25, 30.

**Does anyone know what he counted by? No? He counted by 5's. Anyone else do that? OK, is there another way to think about what's missing?**

Yanni: I counted to the numbers you took out.

**Where did you start counting?**

Yanni: At 11.

**Let's try it.** *[The class counts from 11 to 15 by 1's.]*

**Did anyone else count?**

Claire: I did it sort of like that. I did it from 1, 2, 3, 4.

**So, Claire thought about it in even a different way. She counted up from 1. Any other ways? We had counting, patterns, and thinking about what comes before. What a lot of ways to find just one number!**

# Which Holds More?

## Materials

- Empty containers, labeled with letters
- Interlocking cubes (class set, at least 60 per pair)
- Large chart version of Student Sheet 12, Which Holds More?
- Student Sheet 12 (1 per student)
- Hundred Number Wall Chart
- Plastic Hundred Number Boards with removable number tiles
- Adding machine tape cut in 2-foot strips (from Sessions 1–2)
- Clear tape (available)
- 100 chart (2 or 3 per class, optional)
- Calculators (at least 6–8 per class)
- Game Record Sheet (1 per student, homework, optional)

## What Happens

Students continue Choice Time after the introduction of two new choices: Which Holds More? and Exploring Calculators. In Which Holds More? students estimate which of two containers holds more cubes, and check their estimates by filling them and counting the cubes. Students' work focuses on:

- counting quantities to about 40
- using numerals to record how many
- estimating more and less volume
- reading, writing, and ordering numbers to 100
- finding the larger of two quantities, up to about 40
- exploring patterns in the number sequence
- exploring calculators as a mathematical tool

**Homework Review:** At the start of the session or at another time during the day, you might take a few minutes for students to share their homework. Ask several students to tell the number they found at home, and to explain where they found it.

## Activity

### Which Holds More?

Which Holds More? is a rich activity that touches on several areas of mathematics. In this unit, the emphasis of this activity is on counting and comparing, while in the Measuring unit *Bigger, Taller, Heavier, Smaller,* students do a similar activity to explore capacity.

As you introduce this activity, encourage students to make and explain predictions about which container holds more. This helps students begin exploring some of the ideas they will focus on later in the year, and also highlights the importance of making and checking predictions. The **Dialogue Box,** Which One Holds More? (p. 98), shows how one teacher introduced the activity.

Bring a large container of interlocking cubes and several of the labeled containers to the meeting area. Encourage students to share a few things they notice about the containers (their size, their shape, the letters on them, what they might be used for).

Explain that during Choice Time, the students will be looking at two containers at a time to decide which one holds more cubes. Hold up two containers, asking students to predict which is larger and to explain their reasoning. You might choose two that have very different shapes but are fairly close in number of cubes they hold (for example, one that holds about 20 and one that holds about 30). When students have offered several predictions, ask for ideas on how they might find out for sure which is larger. If no one suggests using the cubes, do so yourself:

**Let's fill both containers with cubes to see which holds more. About how many do you think each would hold?**

Fill the containers with cubes. Then ask students to help you count as you remove cubes from one container and snap them together. Do the same with the other container. Encourage students to revisit their predictions about which container held more. Are they surprised at which held more? at how many cubes each held?

Post your large chart version of Student Sheet 12, Which Holds More? Use this to show students how to record their work during Choice Time. Demonstrate filling in the letters to show which containers you used. Then, ask students for ideas on how to show which container holds more and how they know.

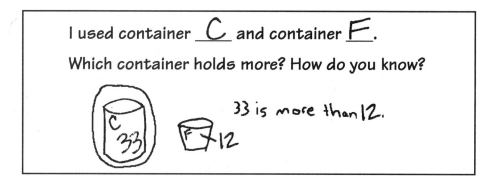

Student suggestions might include writing the numbers of cubes and marking the larger container; drawing pictures of all the cubes; writing in words how many cubes they used for each, and explaining that the one with the larger number holds more. Model several students' suggestions. Explain that whatever they write should include the number of cubes they counted for each container. So, for example, if they draw a picture of the cubes, they need also to write the number.

## Exploring Calculators

Take a few minutes for students to share some experiences they have had with calculators this year, or ways they have seen calculators being used. Explain that during Choice Time, they will have a chance to explore calculators further. They might make numbers on the calculator, try to figure out what some of the different keys do, or make up problems to solve.

**Note:** If your students have not used calculators yet this year, or if they have not used them in several months, you might want to briefly review or introduce guidelines for caring for and storing them. See the **Teacher Note,** Using the Calculator in First Grade (p. 171), for more information.

## Choice Time

Choices

1. Missing Numbers

2. Counting Strips

3. Which Holds More?

4. Exploring Calculators

Add the two new activities to the posted Choice Time list from Sessions 1 and 2. Students may work on any of the four choices. While it is important to give students the opportunity to freely explore calculators throughout the year, they need spend only 10–15 minutes on their exploration each time. Each student should have completed a counting strip by the end of Session 4.

See pp. 84–86 to review Choices 1 and 2, Missing Numbers and Counting Strips.

## Choice 3: Which Holds More?

**Materials:** Containers labeled with letters; interlocking cubes (class set); Student Sheet 12, Which Holds More? (1 per student)

Students can work in pairs or individually, choosing two containers to compare. They write the letters of the chosen containers on Student Sheet 12, then fill both containers with cubes. You may need to discuss what "full" is if students seem to be piling cubes on. A simple rule is that cubes can't protrude above the rim of the container. To help with the comparison, students might fill each container with cubes of a different color (for example, all red in Container A and all blue in Container B). Students then count and compare the number of cubes in each container to see which one holds more, and record their results, being sure to include the *number* of cubes each container held.

Encourage students to include comments about their result (you may want to jot these down on their paper for them); for example:

I knew this would hold more because it's bigger up and down and it's bigger the other way, too.

I thought C would hold more than A because it's so tall, but it didn't.

**Note:** If students do not have enough cubes to fill both containers at the same time, they can fill one, count and record the number of cubes it held, and then use the same cubes to fill the other.

**Different Levels of Challenge**  If some students seem to be having a great deal of difficulty counting accurately, suggest that they use smaller containers, such as those holding between about 10 and 20 cubes.

For more challenge, suggest that students put three or four containers in order, from the one they think will hold the least cubes to the one they think will hold the most. They record the order, then they fill the containers with cubes, count, and compare.

## Choice 4: Exploring Calculators

**Materials:** Calculators

Students spend some time exploring a calculator. Encourage students to share any discoveries they make about calculators with one another.

Your students may vary widely in what they notice about or do with a calculator. Some may use calculators only to make numbers; some may use calculators to explore patterns, such as those created by repeatedly adding 2's; and some may make up and solve problems with calculators.

Keep in mind that some students who can create and solve problems with calculators (or even without calculators) may have little understanding of what the operations they use mean, and little or no sense of whether the answer they obtain is reasonable. Do not insist that students work only with operations they understand well or with small numbers. Often when students are exploring calculators, they become fascinated by the power of being able to make large numbers or solve complicated arithmetic problems, even if they don't completely understand these.

## Observing the Students

See pp. 88–89 for guidelines on observing students' work on Choice 1, Missing Numbers, and Choice 2, Counting Strips.

### Which Holds More?

- Do students count cubes accurately? Do they coordinate one number name with one cube as they count?

- Do students connect the cubes they have counted or use some other method to make sure they have counted each cube once and only once?

- Do students say the counting words in the correct sequence? What parts of the oral counting sequence are difficult for them? Is the order of the numbers between 10 and 20 particularly difficult for some students? Do students know what comes after 19 or 29?

Keep in mind that some students may be able to say the number sequence much further than they can count actual quantities. For example, they may count aloud to 50 quickly and easily, but may not be able to count 25 objects accurately. Review the **Teacher Note,** Observing Students Counting (p. 23), for more on the development of counting abilities in first grade.

### Exploring Calculators

- Do students know how to turn the calculator on and off? Do they recognize that the digits they enter appear on the screen display? Do they know how to clear the screen?

- Are they familiar with the +, –, and = symbols? with any other symbols on the keys?

- What do students do with the calculator? Do they make numbers? explore patterns? create and solve computation problems?

- What numbers can they read? If a two-digit number appears on the screen display as a result of a computation, can they tell you what the number is? Do they know the names of any three-digit numbers that appear on the screen display?

**Note:** Students may encounter decimals during their explorations of the calculator. If some students want to know what "the little dot" means, ask them first for their ideas.

Many first graders will not yet be ready to interpret the decimal portion as part of a whole number. Explain to them that they'll learn more about numbers with dots, or decimal points, in the next year or two. For now, when they encounter a number like 5.326, they can think of it as "about 5," or "5 and a little more," or "5 and some extra."

**Looking at the Counting Strips**  At the end of Session 4, collect students' counting strips. Look at them to collect common errors (errors that several students have made), such as reversing digits or skipping a decade. You will focus attention on these in a later activity (Discussing Our Counting Strips, p. 104).

Some students may become intrigued with writing higher and higher numbers and want to keep working on their counting strips over the next few days or weeks. Return their counting strips after you have reviewed them and encourage them to continue working when they have free time.

## Sessions 3 and 4 Follow-Up

**Math Games**  Students will benefit from repeating any of the games they learned in this unit, playing with someone at home. You may want to send home a copy of the Game Record Sheet, either specifying one game or giving students a choice among Compare Dots, Copying Counters, Double Compare, On and Off, and Counters in a Cup.

**Homework**

# Which One Holds More?

To introduce the activity Which Holds More? this teacher gathers students around the collection of containers and asks students to share their observations about them. Throughout the discussion, she invites students to make predictions and to explain their reasoning.

**One of your choices during Choice Time is to pick two of these containers and find out which holds more.** *[The teacher picks out container C, a large green plastic drinking cup, and container H, a large margarine tub.]* **How about predictions about which container might hold more?**

**Nadia:** The green cup [C].

**Why?**

**Nadia:** It's big.

**Which way is it big?**

**Nadia:** From the top. And down to the bottom.

**Shavonne: I think H because it's wider, really wide.**

**Max:** I think both. Both will hold the same.

**Why? How can that be when they look so different?**

**Max:** Because one is larger and one is fat, but they both equal the same, because large and fat don't make a difference, it's how big.

**Susanna:** The C one because it's very tall and a lot of those cubes will go in it.

**The cubes! That sounds like an excellent idea for how we might try to figure it out. First let's make an estimate. How many cubes do you think these containers would hold?**

**Garrett:** Ten for H.

**If H holds 10, how many do you think C will hold?**

**Garrett:** Sixteen.

**So you think C will hold more than H.**

**Shavonne:** I was going to say the opposite of Garrett. C will hold 10 and H will hold 16.

**Susanna:** Nine, no... 11 in H and 17 in C.

**Max, you said they would hold the same amount. How many do you think?**

**Max:** I say 41 in each.

*[Students call out that this isn't possible, it's much too big a number.]*

**Let's see. Let's try filling them. We're going to fill the containers to the top and then pour them out and count the cubes. Let's use a different color for each container so we don't get mixed up.**

*[The teacher fills C with red cubes and H with blue cubes. When she gets to the top of each container, she points out that "full" means level across the top.]*

**Let's count them up. Which should we dump first? OK, let's try C first.**

*[The teacher snaps cubes together as the class counts them aloud. They count 23 cubes from container C, and they count again to double-check. Next, they count the cubes from H and get 32.]*

**That's close. So which container held more?**

**Iris:** The H.

**Some people thought C had more. Why do you think H ended up having more?**

**Tony:** It's fat and a little bit big.

**Leah:** It's wide and fat, and not too short.

**Shavonne:** Because I think, because it's kind of, you know how applesauce comes in things like, kind of low and fat, and when you eat applesauce and you think that there's not a lot in there, but there really is?

# More Counting and Comparing

## What Happens

Students gather at the Hundred Number Wall Chart to play a group game of Missing Numbers. Then they learn to play Ten Turns, rolling number cubes and accumulating pennies or counters. For most of these sessions, students continue Choice Time with Missing Numbers, Which Holds More? Exploring Calculators, and Ten Turns. At the end of Session 7, the class discusses common errors in sequences of written numerals on their counting strips. Their work focuses on:

- finding the total of two quantities, one that's just a few, and another up to about 40
- counting quantities up to about 40
- using numerals to record how many
- reading, writing, and sequencing numbers to 100
- exploring patterns in the number sequence
- finding the larger of two quantities, up to about 40
- estimating more and less volume
- exploring calculators as a mathematical tool

### Materials

- Counters, such as plastic pennies (50–60 per pair)
- Number cubes: 1–2 per pair
- Student Sheet 13 (1 per student, homework)
- Student Sheet 14 (1 per pair, plus 1 per student, homework)
- Containers labeled with letters
- Interlocking cubes (class set)
- Student Sheet 12 (copies remaining from Sessions 3–4)
- Hundred Number Wall Chart
- Plastic Hundred Number Boards with removable number tiles (2 or 3)
- 100 chart (2 or 3 per class, optional)
- Calculators (at least 6–8 per class)

## Activity

### Missing Numbers

Gather students together for a few minutes to play Missing Numbers on the Hundred Number Wall Chart. Playing as a whole group from time to time gives you the opportunity to hear students' thinking and reasoning, as well as to get a sense of the class as a whole.

Remove five number cards in a pattern, such as 9, 12, 15, 18, and 21. Have the numeral cards for these numbers and red chart markers handy and use them to fill in the blanks as students decide what the missing numbers are. Be sure to ask how they decided which number goes in each space.

In order to involve more students, try one or both of the following:

- Ask several students to offer their ideas on how they know a particular number goes in the blank. For example, one student might use the number before the blank; another might use the number after the blank; another might notice something about the row or column of the missing number; another might notice a pattern in the numbers filled in so far.
- Give each student an index card or slip of paper and a pencil or crayon. As you point to each blank space, students who think they know the missing number can write it on their paper. After a minute or so, ask all students who have written a number to hold it up. Students who are ready to participate can, while those who are not sure don't need to volunteer a number.

After all the missing numbers have been replaced and covered with red chart markers, ask students what they notice about the red numbers, and how they would extend the pattern for two or so numbers in each direction.

**After 21, which is the next number that should be red? Then which number? What number before 9 should be red?**

Slip red chart markers over these numbers, as students direct you.

## Introducing Ten Turns

Ten Turns is a game in which students accumulate pennies, cubes, or other small counters. As students play this game, they practice connecting the oral counting sequence and written numerals with the quantities they represent.

On the board or chart paper, write four copies of the sentences that appear on the Ten Turns Game Sheet (Student Sheet 14). Briefly demonstrate the game with a volunteer partner and show how to record each turn.

Ask your partner to roll one number cube and take that number of pennies or counters. Read the first pair of sentences with students or have a volunteer read them, and show students how they would fill in the blanks. For example:

I rolled _4_ . Now we have _4_ .

Ask students to predict what might happen if small numbers are rolled:

**It's my turn to roll next. If I roll a 1 next, how many do you think we'll have? How do you know? If I roll a 2 next, how many do you think we'll have? OK, let's see what happens.**

Keep taking turns with your student partner. After each turn, ask students to help you fill in the blanks. You will end up with something like this:

I rolled <u>4</u>. Now we have <u>4</u>.

I rolled <u>2</u>. Now we have <u>6</u>.

I rolled <u>1</u>. Now we have <u>7</u>.

I rolled <u>5</u>. Now we have <u>12</u>.

If you are using pennies, ask students how much money you have so that you can model using the term *cents* (4 cents, 6 cents, 7 cents, 12 cents) and ways to write these amounts (4¢, 6¢, 7¢, 12¢).

---

**Activity**

## Choice Time

Add Ten Turns to the posted list of choices and cross out Counting Strips. (Although students might continue to add more numbers to their counting strips over the next few days, they need to spend Choice Time working on the other activities.)

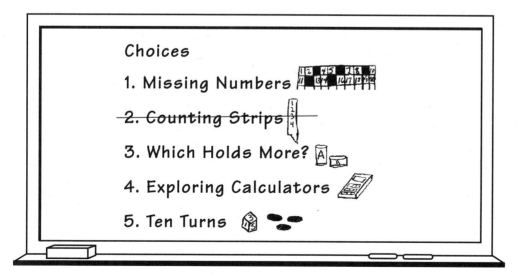

Choices

1. Missing Numbers

2. ~~Counting Strips~~

3. Which Holds More?

4. Exploring Calculators

5. Ten Turns

Students will work on Choice Time for the rest of Session 5, all of Session 6, and the first half of Session 7. By that time, students should have worked on all four remaining choices: Missing Numbers, Which Holds More?, Exploring Calculators, and Ten Turns.

If you have an extra few minutes at the beginning or end of Session 6, play a whole-group round of Missing Numbers at the Hundred Number Wall Chart. Toward the end of Session 7, announce the end of Choice Time and gather students for a discussion of their counting strips.

Refer to Choice Time in Sessions 1–2 (p. 86) and Sessions 3–4 (p. 94) to review Choices 1, 3, and 4.

## Choice 5: Ten Turns

**Materials:** Counters or pennies (50–60 per pair); number cubes (1–2 per pair); Student Sheet 14, Ten Turns Game Sheet (1 per pair)

In this game, players work together to collect counters. On each turn, one player rolls the number cube and takes the number of counters shown. Together, the two players find the total number of counters. Then, the other player records the number rolled and the total number of counters collected. For the next turn, they switch roles. The game is over after ten turns.

**Different Levels of Challenge** Students who need more practice working with numbers in the 20's and 30's can play for fewer turns, such as five. Students ready to work with numbers 50 and above can play for 20 turns. They will need two game sheets (Student Sheet 14) and more counters.

Some students may be ready to use *two* number cubes. On each turn they roll both cubes, find the total shown, and take that number of counters.

## Observing the Students

Refer to pp. 88 and 96 for guidelines on observing student work on Choices 1, 3, and 4.

### Ten Turns

■ Do students count the counters accurately? Do they have ways of keeping track of which objects they have counted? Up to what quantities are students comfortable counting?

To help students who consistently miscount even small quantities, engage a few students in conversation about how easy it is to miscount and what strategies they use to keep track of the objects they are counting. How do they know they've counted every counter? How do they know they won't count a counter twice?

Let students demonstrate strategies they have come up with to count carefully, such as moving each counter into a new pile as they count, or putting the counters into a line or some other organized arrangement so they don't lose track. You might demonstrate a disorganized method of counting yourself (spreading your counters in a disorganized way on the table and counting so that you count many of them more than once), and ask students to suggest ways to help you count.

■ Can students record their quantities accurately with written numerals?

■ What strategies do students use to find the total after each roll? Do they recount the entire set from 1 each time? Do they count on from the previous total? Do they work just with the numbers they recorded on the sheet? Do they "just know" some sums? Can they add on 1 or 2 to some numbers? to any number?

Besides practice with spoken numbers, written numerals, and counting, this activity gives students the opportunity to think about what happens when they add on small numbers of objects to a set of objects they already have. You will probably see a wide range of what students can do. Some students will not be able to add on even 1 or 2 without recounting the entire set; others will begin to reason about what happens when 1 or 2 objects are added; some will count on, starting with the quantity they have. ("I have 13 now, I rolled a 3, let's see, so that's 14, 15, 16.")

Avoid specifically teaching students how to count on. Throughout the grade 1 units, they will have many opportunities to combine small quantities and will begin developing their own strategies for counting on.

## Discussing Our Counting Strips

Gather students at the end of Session 7 for a brief discussion about their counting strips. Write three or four counting sequences on the board or on chart paper, each sequence containing an error that commonly arose in your students' work.

| 9  | 10 | 25 |
|----|----|----|
| 10 | 11 | 26 |
| 11 | 12 | 27 |
| 12 | 13 | 28 |
| 13 | 41 | 29 |
| 14 | 15 | 20 |
| 16 | 16 | 21 |
| 17 | 17 | 22 |
| 18 | 18 | 23 |
| 19 |    | 24 |
|    |    | 25 |

Ask students to discuss what's wrong, how they know it's wrong, and how they could fix it. Encourage students to think of reasons why someone might make the mistake you've illustrated, not just what the mistake is. The **Dialogue Box,** What Went Wrong? (p. 105), shows how students in one class were thinking about reasons for counting errors.

Learning to correctly write numerals in sequence takes time. Many students have difficulty when crossing decades (from 19 to 20, 29 to 30). Others get confused about the order of the digits in the teen numbers. Others may omit numbers in the counting sequence. Students at this age are learning the oral counting sequence, the written counting sequence, and how these written and spoken symbols represent quantities. The process is a complex one. As students count orally, see the written numerals, and use both number names and written numerals as they count objects, they will begin to make more sense of how these expressions of number are coordinated. For more information, review the **Teacher Note,** Observing Students Counting (p. 23).

## Sessions 5, 6, and 7 Follow-Up

### 🏠 Homework

**Ten Turns** Students teach someone at home to play Ten Turns. They will need Student Sheet 13, Ten Turns (the directions), Student Sheet 14, Ten Turns Game Sheet, a number cube, and pennies or other counters (buttons, paper clips). Some students may have a number cube at home as part of a board game. If not, they can use the 1–6 cards from their deck of number cards, sent home in Investigation 2.

# D I A L O G U E   B O X

## What Went Wrong?

The teacher has recorded three sequences of numbers that represent typical errors she saw students make on their counting strips. Although she asks students to correct each sequence, the emphasis of the discussion is on why someone might have made the errors.

| 14 | 31 | 28 |
|----|----|----|
| 15 | 32 | 29 |
| 17 | 33 | 40 |
| 18 | 43 | 41 |
| 19 | 53 | 42 |
|    | 36 | 43 |
|    | 37 |    |

**I've been looking at your counting strips, and I copied on the board some things that went wrong. What could have happened in the first strip?**

Michelle: A number is missing.

**Some people are so excited about writing numbers that they are leaving numbers out. Can someone show us what's missing?**

*[Tuan comes to the board and writes 16 between the 15 and 17.]*

**What happened in the second example?**

Brady: 53 and 43 are wrong.

**What do you think happened?**

Brady: They [the numbers] should be in the 30's.

**What do you think this person was thinking?**

Kaneisha: They went so fast. They put the wrong numbers.

Susanna: In those two *[pointing to 43 and 53]* the 3 should be in front.

Donte: They wrote it the opposite way.

**Who wants to show us how they think it should be?** *[William writes 34 next to the 43, and Donte writes 35 next to the 53.]*

**Let's look at the last example.**

Tony: It should be 38 and 39.

**What do you think happened?**

Mia: They skipped 10 numbers, 30, 31, 32, 33, 34, 35, 36, 37, 38, 39.

Tony: They could have started at 40, and then counted back wrong.

**OK, so Tony thinks they skipped the 30's when they counted back, and Mia thinks they skipped the 30's when they counted forward. What could they think about for next time?**

Mia: It goes 20's then 30's then 40's.

Michelle: If you go higher and don't know your numbers, you don't have to skip them, you can ask a friend or look it up on the number line.

# Clapping Patterns

## Materials

- Interlocking cubes (about 200 available)
- Pattern blocks: 1 bucket
- Hundred Number Wall Chart with red chart markers
- Cube pattern strips (an available supply)
- Blank 100 charts (10–15 available)
- Unlined paper (available)
- Crayons or markers

## What Happens

In Clapping Patterns, an activity that is repeated throughout grade 1, students act out repeating patterns with a set of physical actions. They also show the pattern in one or more other ways: with drawings, with numbers, or with objects like cubes or pattern blocks. Students' work focuses on:

- making and describing repeating patterns
- representing the same pattern in different ways: through physical actions, objects, drawing, and numbers

## Clapping Patterns

If you are doing the full year *Investigations* sequence for grade 1, students will be familiar with Clapping Patterns from the unit *Mathematical Thinking at Grade 1*.

Clapping Patterns involves simple physical actions, such as clapping hands, slapping knees, and putting both hands on head, done in a repeating pattern. For example, the group might repeat together knees-knees-clap or clap-clap-knees-knees. Students then show the pattern in one or more other ways: with materials such as colored cubes or pattern blocks, by drawing, or with numbers.

Briefly review how to do Clapping Patterns, or introduce the activity if your students haven't done it before.

**Warming Up** Gather students in a circle around you, so that everyone can easily see as you lead the patterns. Begin a simple pattern such as knees-knees-clap: slap your knees with your hands twice and then clap your hands. Keep a steady rhythm, with one beat for each knee-slap or clap. Once you have set the pattern, ask students to join in.

After several repetitions, do a few other simple patterns. Include some with four beats, such as knees-knees-clap-clap or clap-clap-clap-knees. You might ask students to suggest patterns. You or a volunteer can begin each pattern, and the class joins in.

**Showing Patterns with Materials**  After this introduction, students will be representing a clapping pattern in another form. Take a few minutes to demonstrate how they might show a clapping pattern with interlocking cubes. Choose a pattern you clapped with four beats and two actions, such as clap-clap-knees-knees. Clap the pattern again with the class two or three times. Then, ask students to describe it to you in words:

**What can you tell me about this pattern? Who has another idea?**

**Let's show the pattern with cubes. I'm going to use blue for clap and white for knees. How could I make a cube tower to show the clap-clap-knees-knees pattern?**

If students don't understand this question, do the clapping pattern with them again. This time, while you are clapping, say the words "blue, blue, white, white" as you act out the pattern. Then ask your question again. As students direct you, build a tower that shows the pattern two or three times: blue, blue, white, white, blue, blue, white, white.

**Do you think we could show clapping patterns on the 100 chart?**

Turn students' attention to the Hundred Number Wall Chart, and have red chart markers available.

**Let's say the black numbers are claps, and we'll make them red for knees. What should I do to show our clap-clap-knees-knees pattern?**

If necessary, demonstrate with the cards up to 8, inserting the chart markers over every third and fourth number. Then point to them, saying "black, black, red, red, black, black, red, red." Ask students to tell you what should come next. Keep going up to about 20 or 30, depending on students' attention and interest, then ask students what they notice about the pattern on the 100 chart.

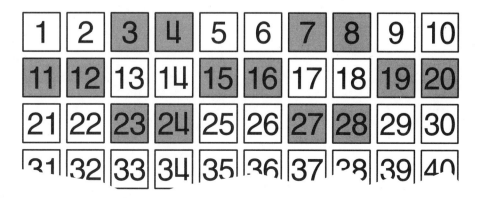

## Representing a Pattern

**We used cubes and we used the 100 chart to show the clap-clap-knees-knees pattern. What else could we use? Who has other ideas? Could you draw a picture to show the same pattern? What could you draw?**

For the rest of the session, students work on their own to show the clap-clap-knees-knees pattern by building or coloring. Make available cubes, pattern blocks, cube pattern strips, blank 100 charts, unlined paper, and crayons or markers.

Every student should end up with a representation of the pattern on paper. Students who build with cubes can record their work on the cube pattern strips, while those who use pattern blocks can trace or draw them to record their pattern. Some students may want to make their own drawings of the pattern. Students who use the blank 100 chart will need help learning how to continue the pattern from the end of one line to the beginning of the next. If this is too difficult for some students who choose the blank 100 chart, you might suggest they switch to the cube pattern strips.

## Observing the Students

As students are working, watch for the following:

- Do students represent the pattern accurately? Do they use repetition in their patterns to help them?

  Some students may find it helpful to act out the pattern as they color the chart.

- Can students explain what a particular part of their representation shows about the pattern? For example, if you point to a particular pattern block or cube in their representation, can they tell you if it stands for a *clap* or a *knee*? If they have built a long train of interlocking cubes to show the pattern, can they identify the unit of four cubes that shows the basic pattern clap-clap-knees-knees?

- What predictions can students make about their patterns? For example, can students coloring a blank 100 chart predict the color of a square in the next row that they haven't filled in yet? Do they base their predictions on patterns in the colored numbers they have written so far? Can students building the pattern from cubes tell you what color cube they'll put on next? what color the next cube after that will be? what color the fourth cube will be?

**Creating New Patterns**  As students finish their representation of the clap-clap-knees-knees pattern, they might make a new pattern with one of the materials. If time permits, gather the class again and ask them how they might clap one or two of the students' patterns, or save these patterns for the next time you do Clapping Patterns.

You may want to make an exhibit of students' representations of the clap-clap-knees-knees pattern.

# Counting Stories

## Materials

- *Out for the Count: A Counting Adventure,* or similar book
- Hundred Number Wall Chart
- Drawing paper
- Crayons or markers

## What Happens

Students listen to a counting story. They estimate and count some of the quantities in the book, and find the corresponding numbers on the Hundred Number Wall Chart. Each student then illustrates one page for a class book of a similar counting adventure. Their work focuses on:

- estimating quantities
- counting quantities up to 100
- sequencing numbers up to 100
- locating numbers on the 100 chart

## Activity

### A Counting Story

Gather students so that everyone can see as you read aloud and show the pictures in *Out for the Count: A Counting Adventure* by Kathryn Cave. This book tells the story of a little boy who can't sleep. He tries to count sheep but instead imagines that he is encountering various other creatures. On each two-page spread, he meets something different: 7 sheep, 12 wolves, 23 pythons, 36 goats, 45 pirates, 54 penguins, 61 bears, 70 bats, 88 ghosts, 97 tigers, and 100 shadows.

Stop on some of the pages, cover up the text with your hand so that only the picture is visible, and ask questions like these:

**Now what does Tom see? Yes, snakes. Does anyone know what kind of snakes they might be?... Diego thinks they might be cobras and Tamika thinks maybe rattlesnakes. We'll find out when we read.**

**How many snakes do you think there are here? Without counting, what can you tell me? Do you think there are more than 10? more than 20? more than 100? OK, let's read and see.**

In the book, each group of animals is shown on the right-hand page grouped into tens and ones. If some of your students notice this, you can count by 10's with them. However, using different units like this (counting by 10's) is a complicated idea that most students continue to work on well into second or third grade. Many students will really trust the number they get only when they count by 1's. However, it is fine for them to hear what counting by 10's sounds like.

When you have finished the book, ask students to make a list with you on chart paper or the board of everything Tom encountered. Show them the appropriate pages of the book as reminders. List the number and kind of creature, like this:

7 sheep
12 wolves
23 pythons
36 goats
45 pirates
54 penguins

Now ask students to help you locate each number on the Hundred Number Wall Chart. Insert the red chart markers over the numbers for each of the creatures, so that you have red numbers for 7, 12, 23, and so forth. Briefly ask students what they notice about the numbers.

**What can you say about the numbers in Tom's story? What happens each time we go to the next page?**

Students may notice that the numbers always increase, and that there is one number for each row on the 100 chart.

**Note:** If you do not have *Out for the Count,* you can adapt this activity to work with another counting book. Some possible alternatives are *One to One Hundred* by Teri Sloat (A Puffin/Unicorn Book, 1991); *Count!* by Denise Fleming (Holt, 1992); or *Animal Numbers* by Bert Kitchen (Dial Books, 1987). Although none of these alternatives has a story element to it, the illustrations are attractive and offer opportunities to count larger quantities. All are suitable models for students' work in the following activity.

## Activity

### A Class Counting Adventure

Explain that the class is going to write their own counting adventure story. Tell students the beginning of a counting story; you can use the one that follows or make up your own.

❖ **Tip for the Linguistically Diverse Classroom** To make the adventure comprehensible to all students, choose volunteers to play the parts of Leo and Maya. Ask them to act out the adventure as you tell it. Also draw simple sketches on the board as new objects are mentioned: cloud, door, pond, turtles on a log.

One day Leo and Maya were walking home from school. When they turned a corner onto the block where they lived, they were surprised to see a big fluffy cloud ahead of them, not in the sky, but on the *sidewalk,* right in front of their house. As they came closer, they saw a door that opened into the cloud. A sign on the door said "Open me." They couldn't resist. They walked right through that door into a strange room.

And in the room was a pond with 5 giant turtles sitting on a log. Behind the pond was another door. A sign on the door said "Welcome." They couldn't resist. They walked right through that door into a strange room. And in the room was ...

Explain that Leo and Maya kept going from room to room, finding different numbers of new creatures in each room. Each student is going to make a page to show one room in Leo and Maya's adventure. They need to choose a number and then draw a picture that shows that number of any creature they want. If they choose something that's hard to draw, they might draw just one or two, and then write below the picture how many of this creature Leo and Maya saw.

Help students choose numbers so that no number is duplicated. For most students, keep the numbers low. You may want to use all the numbers 1 through 15, with a few students choosing higher numbers. You might decide to have pairs or small groups of students work together to draw pictures for larger numbers, such as 25 or 37. If some students do not finish during class time, they may complete their work later in the day.

Four reindeer

Five apple trees

Students can write a sentence that describes their picture, or dictate one for you to write on their page.

When pictures are complete, put them in order by number. Read the whole adventure aloud to the class. At the end of each picture, insert the repetition:

And behind the tree [on the other side of the room, at the top of the ladder, etc.] was another door. A sign on the door said, "Come right in." [Make up your own phrase.] They walked right through that door into a strange room. And in the room was...

Put the pictures together into a book or post them on a bulletin board.

## Session 9 Follow-Up

**Are There 100?** Some students might enjoy trying to count the shadows on the last two-page spread of *Out for the Count*. They are very hard to count, but it is possible to count them by groups. How many bears? How many snakes? How many sheep? Students may discover that there are 10 each of the 10 creatures, including the ones that show only part of the creature.

 **Extension**

# Addition and Subtraction

## What Happens

**Session 1: Introducing Combining Situations**
Students solve combining problems, in which they find the total of two amounts. They record their solution strategies and share them with the class.

**Session 2: Introducing Separating Situations**
Students solve separating problems, in which they find the result when one quantity is removed from another. They again record their solution strategies and share them with the class.

**Sessions 3, 4, and 5: Five-in-a-Row and Story Problems** Students play Five-in-a-Row, a version of bingo, for practice with single-digit addition pairs. The remainder of the three sessions is Choice Time, with the choices Five-in-a-Row and Solving Story Problems. At the end of Choice Time, students gather to share strategies for solving story problems.

**Session 6: Dot Addition** Students play Dot Addition, in which they combine dot cards to make given numbers.

**Sessions 7, 8, and 9: Combining and Separating** Students do Quick Images with combinations of dot cards. They add Dot Addition to their other Choice Time activities, Solving Story Problems and Five-in-a-Row. At the end of Session 8, students share strategies for finding addition combinations in the games. At the end of Choice Time, they again share strategies for solving story problems.

**Session 10: Solving Story Problems** As an assessment, students solve story problems and record their solution strategies. For the remainder of the session, they do Clapping Patterns.

**Routines** Refer to the section About Classroom Routines (pp. 172–179) for suggestions on integrating into the school day regular practice of mathematical skills in counting, exploring data, and understanding time and changes.

## Mathematical Emphasis

- Visualizing combining and separating problem situations
- Developing strategies for solving combining and separating story problems
- Recording strategies for solving combining and separating story problems, using pictures, numbers, words, and equations
- Using dot patterns to model number combinations
- Becoming familiar with single-digit addition pairs

## What to Plan Ahead of Time

### Materials

- Number cubes: 12–16 for the class (Sessions 3–5, 7–9)
- Overhead projector (Sessions 3–5, optional; Sessions 7–9)
- Dot card transparencies, Set A, from Investigation 1 (Sessions 6–9)
- Hundred Number Wall Chart (Session 10)
- Paste or glue sticks (Sessions 3–5, 7–10)
- Counters, such as buttons, bread tabs, or pennies: at least 40 per pair (available)
- Unlined paper (available)
- Chart paper or newsprint (18 by 24 inches): 15–20 sheets (available for use as needed)
- Envelopes for story problems (at least 18)

### Other Preparation

- On two sheets of chart paper, write a combining and a separating story problem. Read through Sessions 1 and 2 first to understand how to choose problems appropriate for your class.
- Duplicate the following student sheets and teaching resources, located at the end of this unit. If you have Student Activity Booklets, copy only items marked with an asterisk.

  #### For Session 1
  Student Sheet 15, Apples and Oranges (p. 209): 1 per student, homework

  #### For Session 2
  Student Sheet 16, Eating Apples (p. 210): 1 per student, homework

  #### For Sessions 3–5
  Student Sheet 17, Five-in-a-Row (p. 211): 1 per student, homework

  Student Sheets 18–20, Five-in-a-Row Boards A–C (pp. 212–214): 1 of each board per pair and a few extras (class), plus 1 of each per student, homework. Prepare a transparency of Board A*.

  Story Problems, Set A (p. 221): 1 per student and 1 extra set*. Cut apart and sort into six envelopes. Paste one copy of the problem on the envelope for identification.

  #### For Session 6
  Student Sheet 21, Dot Addition (p. 215): 1 per student, homework

  Student Sheet 22, Dot Addition Board A (p. 216): 1 per pair and a few extras (class), plus 1 per student, homework

  Dot Addition Cards (p. 220): 1 per pair, plus 1 per student, homework, and 1 transparency*

  Blank Dot Addition Board* (p. 219): several for the class (optional), plus 1 transparency

  #### For Sessions 7, 8, and 9
  Student Sheets 23–24, Dot Addition Boards B and C (pp. 217–218): 1 each per pair and a few extras (class), plus 1 each per student, homework

  Story Problems, Set B (p. 222): 1 per student and 1 extra set. Prepare like Set A.

  Story Problems, Set C (Challenges) (p. 223): enough for about half the class (optional). If using these, prepare like Set A.

  #### For Session 10
  Story Problems, Set D (p. 224): 1 per student. Cut apart and clip in three sets.

# Introducing Combining Situations

## Materials

- Combining problem on chart paper
- Unlined paper
- Student Sheet 15 (1 per student, homework)

## What Happens

Students solve combining problems, in which they find the total of two amounts. They record their solution strategies and share them with the class. Students' work focuses on:

- visualizing what happens in combining situations
- understanding that when two amounts are combined, the result is more than the initial amounts
- developing strategies for solving combining story problems
- recording strategies for combining story problems using pictures, numbers, words, and equations

## Making Sense of Combining

**Note:** In this activity, students solve story problems about combining situations by visualizing the quantities and the result. See the **Teacher Note,** Types of Story Problems (p. 123), for a discussion of combining and other problem types. The numbers in the problems are deliberately kept small so that students can work mentally and can focus on the meaning of the story problem: What happens first in the story? What happens after that? Do you end up with more than when you started? less?

In the next activity, Recording Combining Strategies, students will continue to visualize combining story problems and they will begin the challenging work of learning to record their solution strategies to story problems.

**A Combining Problem** Tell a story problem like the one that follows. (This is *not* the one on chart paper.) You may make up a problem of your own that has a more familiar or timely context for your students, but keep the same numbers and basic structure. See the **Teacher Note,** Creating Your Own Story Problems (p. 132), for other considerations. For this activity, it is important to keep the numbers small. Begin with a number such as 4 or 5, and add only 1 or 2 things.

❖ **Tip for the Linguistically Diverse Classroom** Make story problems comprehensible to all students by acting them out and/or drawing quick sketches of important words (in this case, *bananas*).

Ask students to try to see the story in their minds as you tell it. Some may want to close their eyes to help them concentrate. In order to solve story problems, young students need to be able to think about the sequence of actions in the story.

**Here's a story about something that happened to me the other day at the supermarket. I wanted to buy some fruit. I walked around the fruit and vegetable area in the market until I found the bananas.**

**I picked up one bunch of bananas. I counted them. There were 4 bananas in the bunch. I decided I needed some more. So I picked up 2 more bananas.**

**Now, who can tell me what happened in the story?**

Ask several students to tell what they remember about the story: What happened first? Then what happened? Even if one student tells the story correctly, ask a few more students to tell it as well. Some students may anticipate a question at the end of the story, such as "How many bananas in all did I pick up?" Remind students that for now, you're interested only in what they remember about the story and how they're thinking about it; you're not looking for an answer.

Next, ask students whether there were *more* or *less* than four bananas at the end of the story:

**So, first, I had 4 bananas. At the end of the story, did I have more than 4 or less than 4 bananas? How do you know?**

Do not discourage students from giving their solutions and solution strategies, but keep the focus of the discussion on how they know whether there are more or less than 4 bananas. For example, students might tell what they imagined in their heads when they thought about this situation (such as a bunch of 4 bananas and a smaller bunch of 2 next to it); they might model the problem on their fingers; or they might explain that they know that 4 and 2 more has to be more than just 4.

Try one more story problem with the whole group, such as the one that follows. Again, keep the numbers small and manageable so that students can focus on making sense of what happens when they add on 1 or 2 to a familiar number. Ask students to try to visualize what is happening as you tell the story.

**Pretty soon we ran out of bananas at my house. The next time I was at the store, I bought some more bananas. This time I picked up one bunch that had 6 bananas. Then I picked up 2 more bananas. Try to see this story in your head. Who can tell me what happened?**

Again, ask students to put the story in their own words, and then tell you whether there would be more or less than 6 bananas at the end of the story, and how they know.

## Recording Combining Strategies

Now tell the story of the problem you prepared on chart paper for the class to solve. If you decide to make up your own problem for this activity, keep the same numbers and basic structure as the problem included here. Even though some students may be comfortable finding sums of larger numbers, it is important to keep the numbers small for now while students are focusing on recording how they found their solutions. Later in this investigation, you can adjust the numbers in the problems provided.

❖ **Tip for the Linguistically Diverse Classroom** Continue acting out these stories, linking words like *pencil, table,* and *window* with actual objects in the room.

**I was cleaning up the classroom the other day. I found 5 pencils on the floor under the table. I found 6 more pencils next to the window.**

**Who can tell me what happened in the story?**

Ask two or three students to put the story in their own words. Keep the emphasis on retelling the story, and insist that no one anticipate the answer at this point.

When you are satisfied that students have a good grasp of the story, show them the combining problem you have recorded on chart paper.

I found 5 pencils under the table. I found 6 more pencils by the window. How many pencils did I find?

Distribute unlined paper to each student and explain the task:

**I'd like everyone to find out how many pencils in all. You can solve this problem in whatever way makes sense to you. You can use counters, your fingers, or anything else you need.**

Keep track of how you solved the problem, and write down or draw on paper *how you solved it,* so that someone else can understand exactly what you did. You can use words, pictures, and numbers. When we come back together, we'll share some of your strategies.

Students work individually, but they may discuss their strategies with each other.

Note: The students' task here is not only to solve the problem, but to record in some way how they solved it, using words, pictures, numbers, or a combination. This recording may be difficult for many students, and you will need to insist that they record in a way that helps someone else understand their solutions. As students see a variety of recording methods, they will begin to understand what it means to record their thinking; however, this will take time. See the **Teacher Note,** Writing and Recording (p. 125), for ways to help students. Recording and representing problems and their solutions is a major emphasis of this investigation and will be continued in the unit *Number Games and Story Problems.*

## Observing the Students

Circulate to observe students as they work.

- Can students make sense of the sequence of actions in the problem? If some students are having difficulty, ask them to tell the story in their own words: What happened first? Then what happened? Reread the problem with them if necessary.

- How are students solving the problem? Can they explain their strategies? Do they tell you they "just know" the answer?

  Encourage students to solve this problem in whatever way makes sense to them, including using counters, counting on their fingers, or drawing pictures.

  If students say they "just know" the answer, ask them to find a way to prove it to someone else. For example, if some students record just 5 + 6 = 11, ask them to find a way to prove that 5 and 6 is 11.

  **So, you knew that 6 and 5 is 11, but how could you prove it to someone who didn't know? OK, so you can count up 5 from 6. Can you show me how you'd do that? Now, how could you put that on paper?**

  The emphasis in this activity is on explaining and recording thinking. All students, even those who have already committed 5 + 6 to memory, need to find a way to show a strategy for solving the problem.

- How do students record their work? How clearly can they show their strategies? Do they use numbers? pictures? words? equations?

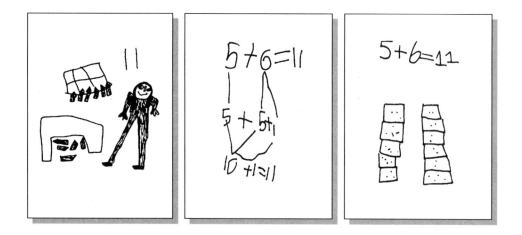

As students show you how they have recorded their thinking, ask them to tell you about how they solved the problem. Compare what they say with what they have written. If there is more to their solution than they have recorded, help them figure out ways to write or draw about what they did.

You'll need to spend a lot of time helping some students understand what it means to record their strategies. The **Dialogue Box,** Recording Strategies for a Combining Problem (p. 126), describes one teacher's interactions with students on this activity. As you observe, make note of some approaches you see, so that you can encourage students with different approaches to share their work with the whole class in the next part of the session.

If you have students who solve this problem easily and record their strategies well, you can ask them to show a second way to find the answer or to solve a related problem with larger numbers. For example:

So, I found 11 pencils. Then I found 9 more under my desk. Now how many did I have?

## Activity

## Sharing Strategies

Gather the group together to share their strategies and how they recorded them. Have counters available for students who want to use them to demonstrate their strategies.

Ask for volunteers to explain how they solved the problem. As students share their approaches, record each different method on the board or on chart paper. You might record each approach using a different color so that students can refer to them easily ("Mine is like the pink way," or "Mine is almost the same as the blue way, but...").

Alternatively, you can number or letter the ways. Avoid labeling them with students' names (Yanni's way, Shavonne's way) because other students are likely to have approached the problem in similar ways, and all students enjoy feeling ownership of their own approaches.

The ways you record will give students models for how they can record their own work. Whenever possible, base your recording on the way the student recorded. If the student drew pictures of pencils, you draw pictures of pencils. If a student used tallies, use them yourself. Sometimes, you will need to develop your own way to show what the student did. Some students may be able to explain the strategies they used, but not know how to put those in writing or not know what is appropriate to record. For example, students who counted on their fingers may have recorded only the numbers they counted. To record this strategy, show a picture of numbered fingers (see example 3).

Use addition notation if students themselves have used it or if you find it a natural way to record a particular strategy. For example, students using a mental strategy, such as knowing that 5 and 5 is 10, and then knowing that 5 and 6 must be 1 more (11), might record the numbers they used but not know how to show what they did with them. To record this strategy, you might use equations (see example 4).

Here are the ways one teacher recorded four different approaches:

**Luis:** I counted out 5, counted out 6, then counted them all.

**Eva:** I started at 5 and counted 6 more.

**Max:** I counted 5 on my fingers, then I counted 6 more on my fingers.

**Mia:** I knew 5 + 5 is 10, so 5 + 6 is one more.

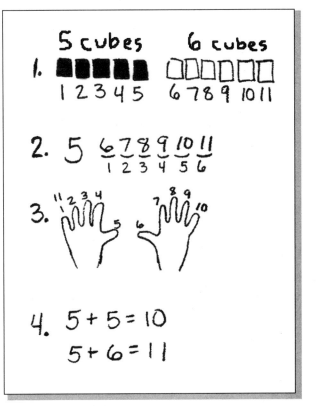

After several students have shared their approaches, ask:

**Does anyone have a way that is different from one of the ways I've written here?**

When you have recorded five or six different strategies, or when students are beginning to become less attentive, ask students to look at their approach and decide which of the ones you've recorded is closest to their own. Ask students to raise hands to indicate which approach they used. This is a way of validating all students' approaches while also giving you a sense of what kinds of strategies are being used in your class as a whole.

If no one in your class has used addition equations and you do not find an opportunity to introduce it as you record students' strategies, model it for the class at the end of this activity. Write $6 + 5 = 11$ on the board or chart paper and ask students if they know what the symbols + and = mean. If necessary, clarify the meaning. Even students who are familiar with addition notation may find it more natural to represent their strategies for solving the combining problem in other ways. Students will have opportunities to use and become familiar with addition and subtraction notation during the rest of this investigation and in the unit *Number Games and Story Problems*. The **Teacher Note,** Introducing Notation (p. 124), further discusses the use of notation in first grade.

# Session 1 Follow-Up

**Homework**

**Apples and Oranges** Send home Student Sheet 15, Apples and Oranges, which gives students another combining problem to solve at home. Remind them that besides the answer, you need to see *how* they figured it out. They should use words, numbers, pictures, or a combination to show their solution strategies.

# Types of Story Problems

The two most familiar types of addition and subtraction story problems are *combining* two quantities to find a total, and *separating*, or removing one quantity from another to find how much is left (what many students call "take away"). In the grade 1 *Investigations* curriculum, students do both. Within these two problem types, there are a variety of problem structures, defined by what information is given and what is to be figured out, as described below.

**Combining** In *joining* or *combining* problems, two or more quantities are combined to form another quantity. These include traditional addition situations:

> Ted had 6 marbles. Sophia gave him 4 more. How many does Ted have now?

In a combining situation like this, the quantities to be combined are known, and the total is to be found. These problems with an *unknown outcome* are probably the most familiar type for your students.

In another type of combining problem, with *unknown change*, the total amount and one of the quantities are known, and the second quantity must be found:

> Ted had 6 marbles. Sophia gave him some. Now Ted has 10 marbles. How many did Sophia give him?

This structure is more challenging for young students than the more typical *unknown outcome* problem.

Note that *how* a problem is solved is not what determines whether it is a combining problem. Some students might solve this problem by a method based on subtraction, such as counting back from 10 to 6; others might use a method based on addition, such as counting up from 6 to 10. However it is solved, the structure is considered combining: Two quantities are combined to create a third. In this case the outcome is known, but the *change* that occurs is not known.

**Separating** In separating problems, one quantity is removed from another, resulting in a por-

tion of the original quantity. These include the familiar "take away" situation:

> Ted had 10 marbles. He gave 4 marbles to Sophia. How many does he have now?

This is a separating problem with an *unknown outcome:* we start with one quantity, change it by removing one or more parts, and find the remaining quantity. As with combining problems, we can alter the structure of the problem by changing what information is given and what we have to find out. In this way we form a separating problem with *unknown change:*

> Ted had 10 marbles. He gave some to Sophia. Now he has 6. How many did he give away?

Separating problems with an unknown change, like combining problems with an unknown change, are more difficult for young students to make sense of.

Students' work with story problems in the *Investigations* curriculum at grade 1 focuses on making sense of and developing strategies for combining and separating problems with unknown outcomes. In this unit, students work with just these two problem types. Students will become familiar with problems with unknown change in the later first grade unit *Number Games and Story Problems*.

Some researchers classify the addition and subtraction problem types described here in slightly different ways. However, this information is all you need to begin recognizing the different structures you will use in providing appropriate problems for your students.

While first graders should become familiar with standard equation notation, it is not essential that they use it themselves at this level. The equation format may seem very straightforward to us as adults, but it actually assumes some complex ideas about number relationships. Second graders are generally more ready to use equation notation to record their work than are first graders. In first grade, many students are grappling with the problem situations and deciding what actions are required to solve them. They are moving from counting by 1's in all situations to sometimes thinking of numbers in larger chunks. They are just becoming familiar with some number combinations. The emphasis for students must remain firmly on making sense of the problem situation and finding a way to solve it. This emphasis includes recording solutions in a way that comes from the student, rather than from the teacher.

As you use equations correctly, students can see how they are used, and some will incorporate them in their own recording. Introduce equations naturally as a way to record student strategies when these strategies are numerical ones. For example, suppose one of your students describes this approach to combining 4 and 8: "I knew that 2 and 2 is 4, so I said 8 and 2 is 10, and then I added the 2 more, and 10 and 2 is 12." This strategy can be nicely recorded with equations:

$$2 + 2 = 4$$
$$8 + 2 = 10$$
$$10 + 2 = 12$$

However, if a student says, "I counted up from 8, so that was 9, 10, 11, 12," this strategy does not suggest equation format. It would be better recorded something like this:

8 ● ● ● ●
  9  10  11  12

Try to match your way of recording as best you can to what the students say, so that a variety of ways of thinking about and recording the problem are valued and shared. At grade 1, it is more important for students to develop clear ways of thinking about the problem for themselves than to use standard notation.

When you do use equations to record, be sure you use them correctly so that students will have a good model. In particular, use *separate* equations to model consecutive steps in a student's thinking:

$$8 + 2 = 10$$
$$10 + 2 = 12$$

It is incorrect to write 8 + 2 = 10 + 2 = 12, which means that the sum of 8 + 2 is equivalent to the sum of 10 + 2. Of course, what a first grader means by this is clear: that she added 8 and 2 first, then added 2 to their sum to get 12. The first grader who writes this way simply understands the notation as a sequence of events, rather than as an equation. Likewise, some first graders will not recognize that "order matters" when using subtraction notation. They might record 2 – 9 to show that they took 2 away from 9. Encourage students to think about whether 2 – 9 means the same thing as 9 – 2.

By the end of second grade, students can be guided to use equation notation correctly, but first grade is too early to do that for most students.

# Writing and Recording

Just as students should be engaged in frequent mathematical conversation, so too should they be encouraged to explain their problem-solving strategies in writing and with pictures and diagrams. Writing about how they solved a problem is a challenging task, but one that is worth the investment of time. As with any writing assignment, many students will need support and encouragement as they begin to find ways of communicating their ideas and thinking on paper, but even the very youngest students can be encouraged to represent their problem-solving strategies.

The range of students' abilities to write and record will vary greatly in any first grade classroom. First graders are just becoming familiar with the areas of reading and writing, and reflecting on one's own thinking is a challenging task in any case. Initially, some students will record just a few words, pictures, or numbers that describe their strategy. Encouraging students to draw pictures as part of their explanation is often a way into the task for many students. Explain to students that mathematicians often write and draw about their ideas as a way of explaining to others how they are thinking about a problem. The more often students are encouraged (and expected) to write and record their ideas, the more comfortable and fluent they become.

Students benefit tremendously from discussing their ideas before writing about them. Sometimes this might happen in pairs, other times in whole-group discussions. Questions and prompts such as "How did you solve the problem?" or "Can you tell me what you did after you put the cubes into groups of five?" may help extend students' thinking. During whole-class discussion, you will model writing and recording strategies for students so they can see ways to record their mathematical strategies using words, numbers, and pictures. For example, when one boy reported to the class how he solved a problem about combining 6 and 5, he said: "I counted out 5 cubes. I counted out 6 more. Then I counted the cubes, and it stopped at 11."

The teacher recorded his strategy on the board as follows, reiterating his strategy in words as she recorded:

I counted out 5 cubes.      I counted out 6 cubes.

■ ■ ■ ■ ■               ☐ ☐ ☐ ☐ ☐ ☐

1  2  3  4  5            6  7  8  9  10  11

I counted all the cubes.

As with any type of writing, providing feedback to students is an important part of the process. As the audience for your students' work, you can point out those ideas that clearly convey their thinking and those that need more detail. If students read their work aloud to you or to a classmate, quite often they can identify by themselves ideas that are unclear and parts that are incomplete. See the **Dialogue Box,** Recording Strategies for a Combining Problem (p. 126), for one teacher's interactions with students.

## Recording Strategies for a Combining Problem

Students are working on the following problem:

> I cleaned the classroom yesterday. I found 5 pencils under Tony's table. I found 6 pencils under Libby's table. How many pencils did I find?

The teacher circulates to help them describe their strategies clearly and to help them find ways to record their work. The three interactions included here illustrate common difficulties students had in approaching the task of recording, and ways that the teacher helped students.

### Drawing Pictures

After writing 5 + 6, Susanna is looking at her paper and seems unsure of what to do next.

**Can you tell me about what you have on your paper?**

**Susanna:** It stands for five plus six... Um, you do 5 plus 6 to get the answer.

**Tell me about what we're trying to find. What's the problem about?**

**Susanna:** Five pencils. And six pencils.

**And what happens to the pencils?**

**Susanna:** You find them... Um, you want to know how many you have.

**So, how could you show that? How could you show what the problem is about?**

**Susanna:** You could draw pencils.

**How many pencils would you draw?**

**Susanna:** Five. And six.

Susanna draws a group of five pencils and a group of six pencils. She thinks for a moment and then says quietly to herself, "Count the pencils." As she counts the pencils out loud, she writes a number below each one. Then she records = 11 after 5 + 6.

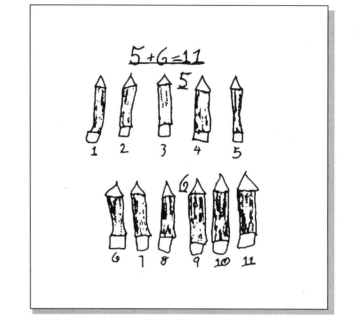

Although Susanna was able to quickly "translate" the story problem into an arithmetic expression, she did not have a way of solving the problem readily. The teacher encouraged her to find a way to model the problem and to use the model to find a solution. For Susanna, the process of recording served both to give her an image of the problem and to give her a strategy for solving it.

### How to Record Counting

When the teacher visits Nadia, she's written 11 on her paper. She tells the teacher that she found her answer by counting.

**Tell me what you counted.**

**Nadia:** I went six and then I, like, counted up.

**How did you know how much to count up?**

**Nadia:** Five numbers, because 6 and 5. Take biggest number to start with, and then count up the rest.

**Well, that's a great strategy. Is there a way you could put that down on paper?**

*[Nadia glances around at the papers of the students sitting nearby, and sees that they are drawing pictures of pencils or of cubes that they used to model the problem.]*

**Nadia:** But, how can I draw a picture of counting?

**What if you just wrote down what you told me you did to solve the problem?**

**Nadia:** Six and then count up 5?

**Yes, you could just write that down.**

Nadia sets to work describing her strategy.

### If You "Just Know" the Answer

Chris has recorded 11. He tells the teacher that the problem was easy, because he knew by heart that 6 plus 5 is 11.

**So, you just knew it in your mind. How could someone who doesn't know it in their mind figure it out?**

**Chris:** They could just memorize it. Or know it by heart.

**Well, let's say they didn't know it by heart. How could you teach them?**

**Chris:** Um… Count on your hands?

**That's an excellent way. Tell me what you'd show them.**

**Chris:** Count 5 *[with the fingers on his left hand]*, then count 6 more *[counts to 10 on the right hand]* and land on 11 *[left thumb]*. Oh, I have another way, too. They could know 5 plus 5 is 10.

**How would knowing 5 plus 5 is 10 help them?**

**Chris:** Six is 1 more than 5, and 1 more is 11.

**Great! So you have two different ways to solve the problem. Try to write down what you did for both of them.**

> 11 because I cattid
> on my fegrs 5 and 6
> and I land on 11
> olso I now that 5+5 is 10
> I mor is 11

# Introducing Separating Situations

## What Happens

Students solve separating problems, in which they find the result when one quantity is removed from another. They again record their solution strategies and share them with the class. Students' work focuses on:

- visualizing what happens in separating situations
- understanding that when one quantity is removed from another, the result is less than the initial amount
- developing strategies for solving separating story problems
- recording strategies for separating story problems using pictures, numbers, words, and equations

### Materials

- Separating problem on chart paper
- Unlined paper
- Student Sheet 16 (1 per student, homework)

## Making Sense of Separating

This session is similar to Session 1, except that students will be solving *separating* problems. The focus of this first activity is helping students visualize what happens when one quantity is subtracted from another: Do you have more or less than when you started? Again, the numbers are kept small so that students can work mentally and focus on the meaning of the story problem.

Tell a story problem to the whole class (not the chart paper problem; you will present that in the next activity). Ask students to listen carefully and try to see what is happening in the story. Some may want to close their eyes as they try to visualize it.

❖ **Tip for the Linguistically Diverse Classroom** As before, make stories comprehensible by acting them out and making quick sketches as a visual clue to important words.

**Here's another story about what happened to me at the supermarket. I was getting tired of bananas, so I decided to buy some apples. I counted out 6 apples into a plastic bag. Then I decided that I had too many. So I put 2 apples back. How many apples did I buy?**

As you did in Session 1, ask a few students to tell the story back to you. Then, ask whether there were *more* or *less* than six apples at the end of the story.

If you think students need more practice making sense of separating, try one more story problem with the whole group.

**While I was walking out to my car, I dropped my bag of 4 apples, and 3 apples rolled away under someone's car. How many apples were left in my bag?**

Again, ask students to put the story in their own words, and then tell you whether there would be more or less than 4 at the end of the story and how they know.

## Recording Separating Strategies

Tell the story of the problem you prepared on chart paper for the class to solve. For example:

**Yesterday morning, I looked out my window and counted 12 squirrels on the ground in my yard. When I went outside, 4 of them ran up a tree.**

Ask two or three students to put the story in their own words. Keep the emphasis on retelling the story, and insist that no one anticipate the answer. When you are satisfied that students have a good grasp of the story, show them the problem as it is written on chart paper.

There were 12 squirrels on the ground. Then 4 of them ran up a tree. How many stayed on the ground?

Distribute unlined paper. Be sure students understand that they must write more than the answer; they must show how they found their solutions.

Most students will see this problem as a subtraction situation and "take away" 4 from 12, or count down 4 from 12. A few students may see this as an "adding on" situation and solve the problem by counting up from 4 or by using knowledge of number combinations. See the **Teacher Note,** Three Approaches to Story Problems (p. 133), for a description of typical student approaches. Another **Teacher Note,** The Relationship Between Addition and Subtraction (p. 136), explains how many problems can be seen in different ways.

If some students solve this problem easily and record their strategies well, they might show a second way to find the answer or solve a related problem with larger numbers:

Later, I saw 15 squirrels in my yard. Then 8 ran away. How many were left?

## Sharing Separating Strategies

As you gather the class for sharing their work, have counters available for students to demonstrate their strategies. Ask for several volunteers to explain how they solved the problem. Record each different method on the board or on chart paper. Whenever possible, base your recording on the student's own approach.

Use subtraction notation if students themselves have used it or if it is a natural way to record a strategy. If some students use notation incorrectly ($4 - 12 = 8$ instead of $12 - 4 = 8$), model the correct way on the board. The **Teacher Note,** Introducing Notation (p. 124), further discusses students' incorrect use of notation.

Here are the ways one teacher recorded five different approaches:

**Brady:** I drew 12 pencils, crossed out 4, counted what was not crossed out.

**Claire:** I put out 12 cubes, took away 4, and counted what I had left.

**Jamaar:** I counted back from 12 and kept track of how many I counted.

**Libby:** I knew that 12 take away 2 is 10, and then take away 2 more is 8.

**Yukiko:** I knew that 4 and 8 is 12, so if you have 12 and take away 4, you have 8 left.

Ask students to look at their papers and decide which of the strategies you've recorded is closest to their own. Ask for a show of hands for each approach.

If you have not yet used subtraction equations, write 12 – 4 = 8 and ask students what they think the minus (–) symbol means. Students may know a lot of different words for this symbol, including *minus, subtract,* and *take away.* Use the first two yourself from time to time so that students hear both *minus* and *subtract.* While *take away* is not a mathematical term, its use by students need not be an issue.

Collect students' work and review it along with last night's homework, to see if students are beginning to represent their strategies clearly.

## Session 2 Follow-Up

**Eating Apples**  Send home Student Sheet 16, Eating Apples, to give students a separating problem to solve at home. Remind them that writing the answer is not enough; they need to show their solution strategies with words, numbers, or pictures.

 **Homework**

The story problems provided in this unit are based on combining and separating situations. Many teachers like to create their own problems; this enables them to use story contexts that reflect the interests, knowledge, and environment of their own students, as well as to adjust the numbers appropriately. We expect and encourage this; however, it is important to provide a variety of problem structures. Whenever you change the contexts or numbers in a problem from this investigation, avoid altering its underlying structure.

**Creating Interesting Contexts** Use contexts that are interesting to students without being distracting. Teachers find that simple situations, familiar to all their students, are the most satisfying. One source of good situations is experiences that you know all your students have had. For example, one urban class walks to a nearby park every day for recess; a problem based on that experience might be:

Today at the park, I counted 6 squirrels on the ground and 2 more in a tree. How many squirrels did I see in the park?

Another class may be having bake sales at lunchtime to raise money for a class trip:

On Friday we sold 6 chocolate cupcakes and 2 vanilla cupcakes. How many cupcakes did we sell?

In one classroom the teacher made up two characters like her students, Ted and Sophia, and built problem situations around them:

Ted and Sophia got a bag of peanuts. As they walked home, Ted ate 6 peanuts and Sophia ate 2 peanuts. How many peanuts did they eat?

Sophia and Ted went to the post office. Ted bought 2 stamps and Sophia bought 6 stamps. How many stamps did they buy?

Many teachers also take advantage of special events, classroom happenings, seasons, or holidays for problem contexts:

Kira made 6 snowballs and Jake made 2. How many snowballs did they have?

**Related Problems** Simple, familiar situations often suggest other problems that easily follow from the initial one. Such follow-up questions might be provided to everyone as options, or only to students who have finished the first problem. Additional problems related to the previous examples might include these:

Yesterday I counted 7 squirrels on the ground and 3 in the trees. Did I see more squirrels yesterday or today?

If we charged 2¢ for each cupcake, how much did we get for all the cupcakes?

If we sold 10 more cupcakes, how many cupcakes would we sell in all?

If Kira and Jake want to have 11 snowballs, how many more do they need to make?

When creating follow-up problems, consider whether to keep the level of challenge the same or to make the problem easier or more difficult.

**Adjusting Numbers in the Problem** You will probably have students who are comfortable working with numbers in the upper teens and higher, and other students who need to work with smaller numbers. In some classrooms, teachers choose two sets of numbers for each problem. For example:

Sophia has 5 pennies. She earned 6 cents. How many pennies does she have now?

Sophia has 12 pennies. She earned 7 cents. How many pennies does she have now?

Understanding addition and subtraction at this level involves making sense of story problems, having strategies for solving them, and being able to communicate strategies orally and in writing. While you may want to challenge students by giving them larger numbers, you can also challenge them by asking them to find answers in more than one way and to find clearer ways to explain their work.

# Three Approaches to Story Problems

Students commonly take one of three approaches to solving a story problem: direct modeling, counting strategies, and numerical reasoning. Each approach is described below and illustrated with examples of student work on this separating problem:

> Last night I picked up 12 pencils from the floor. I put 4 of the pencils in the pencil box. How many pencils did I have left in my hand?

### Direct Modeling

When young students first encounter story problem situations, they usually model the actions in the problem step by step in order to solve it.

Students who are using a direct modeling strategy might count out 12 cubes, then take 4 of them away to represent the 4 that were put in the pencil box, then count the number of remaining cubes.

Leah drew twelve pencils, crossed out four, and then counted the remaining set.

Fernando counted out twelve cubes, took away four, and counted the remaining cubes.

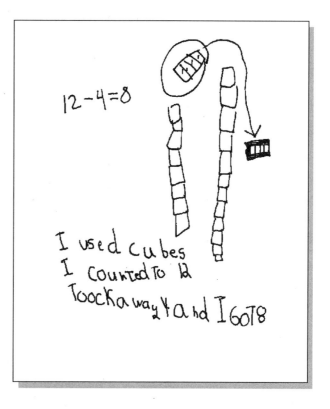

As students gain skill in visualizing problem situations and begin to develop a repertoire of number relationships they know, they gradually develop other strategies based on counting and on numerical reasoning. These strategies require visualizing all the quantities of the problem and their relationships, and recognizing which quantities you know and which you need to find.

### Counting On or Counting Down

Some students, who perhaps feel more confident visualizing the problem mentally, use strategies that involve counting on or counting down.

Donte counted on his fingers. To get 12, he explained that he used both hands and visualized two "imaginary fingers." He counted down from 12, first counting down 2 in his head, using his imaginary fingers, then counting down 2 more on his actual fingers, and got 8.

*Continued on next page*

Donte recorded his counting down strategy on paper like this:

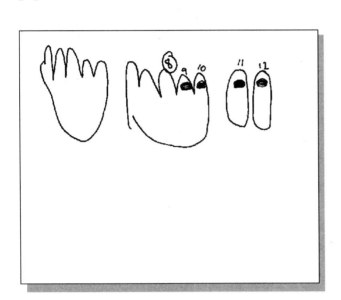

Kristi Ann used the class number line. She started at 12, and counted back 4.

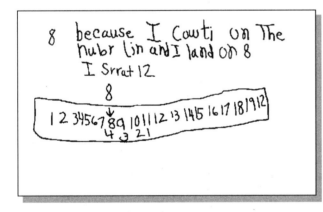

Andre counted down 4 from 12 in his head.

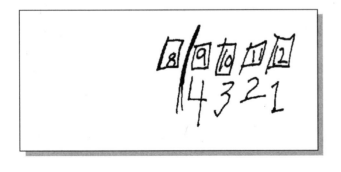

While these three students' methods somewhat resemble the methods of the students that directly modeled the action in the problem, there is an important difference: None of these students had to construct the 12 from the beginning, by 1's. Donte quickly made the 12 out of larger chunks (5 + 5 + 2), while Kristi Ann and Andre simply started with 12. Counting back for subtraction requires a complex double counting method. These students must simultaneously keep track of the numbers they are counting down (11, 10, 9, 8) and the number of numbers counted (1, 2, 3, 4).

Iris used a different counting strategy. She counted out four cubes in a row, to represent the four pencils taken away. She continued putting cubes in a second row, counting on from four until she had a total of 12. Then, she counted the number of cubes in the second row. Iris was able to transform the problem into a different structure, then count on to find the solution: 4 + ___ = 12.

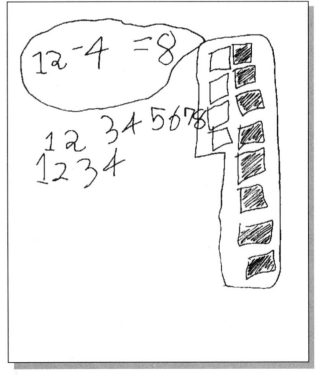

## Using Numerical Reasoning

As students learn more about number relationships, they begin to be able to solve problems by taking numbers apart into useful chunks, manipulating those chunks, and then putting them back together.

Luis broke 4 into 2 and 2. He then subtracted each chunk separately: 12 – 2 is 10, then 10 – 2 is 8.

I nae thet 12–2=10 I take away 2 because it is sapps to be 4=8

Tamika explained, "I know 4 and 4 and 4 is 12. Two 4's is 8, and then there's 4 in the pencil box."

These students are using strategies that involve chunking numbers in different ways, rather than counting by 1's. They are able to visualize the structure of the problem as a whole in order to identify number relationships they know that might help them solve the problem. While it is important to encourage strategies such as these, keep in mind that ability to work with chunks greater than 1 develops gradually over the early elementary years. Many first graders will need to continue counting by 1's for most problems. As they build their understanding of number combinations and number relationships over the next year or two, as well as their ability to visualize the structure of a problem as a whole, they will begin developing more flexible strategies.

# The Relationship Between Addition and Subtraction

It is easy for adults to consider two problem situations as the same, because we would solve them in the same way, while they may actually appear quite different to your students. We may assume that certain situations are addition and others subtraction because we are used to thinking of them that way, but we may find that students solve problems in unexpected ways.

For this reason, as you introduce addition and subtraction problems to students, avoid labeling them as one or the other. A critical skill in solving problems is deciding what operation is needed. Further, many problems can be solved in a variety of ways, and students need to choose operations that make sense to them for each situation. For example, students may solve problems that you think of as subtraction by using addition (in fact, many adults also do this). Consider the following problem:

> 12 squirrels were on the ground. Then 4 of them ran up a tree. How many stayed on the ground?

Most of us learned to interpret this situation as subtraction, and we may naturally assume that students should also see it as subtraction. Students who use direct modeling of the actions to solve the problem will probably count out 12, remove 4, and count how many remain. However, there are many other ways to solve this problem:

- counting down 4 from 12 (11, 10, 9, 8)
- counting up from 4 (5, 6, 7, 8, ...) and keeping track of how many numbers are counted
- using knowledge of number combinations and relationships ("I know 4 + 4 + 4 is 12, and two 4's is 8," or "12 take away 2 is 10, so just take away 2 more, that's 8.")

Some of these methods are based on subtraction (moving from 12 down to 4), but others are based on addition (moving up from 4 to 12). The method chosen depends on a person's mental model of the situation: Do you see this problem as a taking-away situation to be solved by subtraction, as an adding-on situation, or as a gap between two numbers that might be solved by either addition or subtraction, depending on which is easier in the particular situation? Any of these methods are appropriate for solving this problem. Addition is just as appropriate for solving this problem as subtraction, and, for many students, makes more sense.

4 Sqirls in tree
5 6 7 8 9 10 11 12
8 sqirls on grond

# Five-in-a-Row and Story Problems

## What Happens

Students play Five-in-a-Row, a version of bingo, for practice with single-digit addition pairs. The remainder of the three sessions is Choice Time, with the choices Five-in-a-Row and Solving Story Problems. At the end of Choice Time, students gather to share strategies for solving story problems. Their work focuses on:

■ becoming familiar with single-digit addition pairs

■ developing strategies for solving combining and separating story problems

■ recording strategies for solving combining and separating story problems using pictures, numbers, words, and equations

### Materials

■ Student Sheet 17 (1 per student, homework)

■ Student Sheets 18–20 (1 of each per pair, and a few extras, plus 1 of each per student, homework)

■ Overhead projector

■ Transparency of Five-in-a-Row Board A

■ Number cubes (12–16 for the class)

■ Counters such as buttons or pennies (about 20 per pair)

■ Story Problems, Set A (in prepared envelopes)

■ Unlined paper

■ Paste or glue sticks

## Activity

Introduce Five-in-a-Row by playing a demonstration game. You will need an overhead projector, a gameboard transparency (or a chart-paper version for display), and two number cubes.

Roll the number cubes and write the numbers you rolled on the board or on chart paper. Ask students to find their sum and to explain how they found it.

**I rolled a 2 and a 5. What's the sum of 2 and 5? How do you know? Who has another way they know?**

Place a counter on the gameboard on any square that matches the sum. You might ask volunteers to place each counter for you. (If you are using a chart-paper gameboard, draw an X in the box to show where the counter is placed.)

### Five-in-a-Row

| 2 | 3 | 4 | 5 | 6 |
|---|---|---|---|---|
| 6 | 7 | 7 | 8 | 9 |
| 10 | 11 | 12 | 11 | 10 |
| 9 | 8 | 7 | 7 | 6 |
| 6 | 5 | 4 | 3 | 2 |

Roll the number cubes again, record your rolls, and ask students to find their sum. Before placing the next counter, explain that the goal of the game is to mark five squares in a row horizontally, vertically, or diagonally. Encourage students to think about where they could place the counter to help them get five in a row.

Continue playing for a few more rolls of the number cubes, or until you think everyone understands how to play. Make sure students understand that they can cover only one square on each turn; when there are duplicate numbers on the board, they have a choice. If no move can be made (for example, the sum of the cards is 7 but all the 7's are covered), roll again.

Explain that students will be playing this game in pairs during Choice Time, working together to get five in a row on one board.

## Activity

### Choice Time

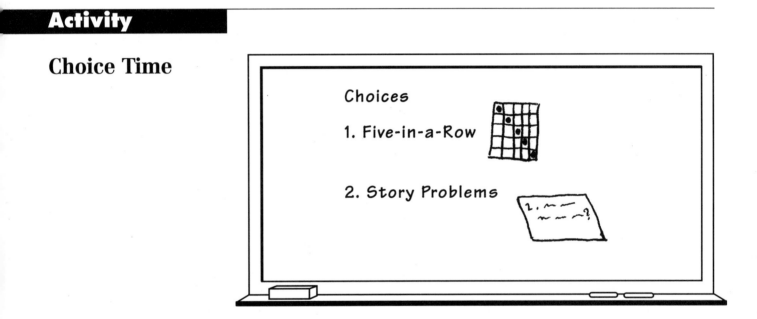

Announce Choice Time for the rest of today (Session 3) and most of next two sessions. Post the two choices.

Since Five-in-a-Row will be fresh in students' minds, you may want to spend more time helping groups of students get started on story problems.

Specify two problems that students are to have done by the middle of Session 5, in time for a class discussion. Choose, for example, problems 2 and 4, or another two more appropriate for your class.

## Choice 1: Five-in-a-Row

**Materials:** Student Sheets 18–20, Five-in-a-Row Boards A–C; counters, cubes, or coins (about 20 per pair); number cubes (2 per pair)

Students play cooperatively in pairs; three may play together if necessary. Each pair needs a gameboard, two number cubes, and 20 counters for marking the boards. (Players may also use counters to help them find sums.) The object of this game is to get five markers in a row, either horizontally, vertically, or diagonally, by covering the sum of two numbers rolled. Students may trade boards with other pairs between games, so they play on different boards.

For more challenge, you might introduce one of the game variations (see Student Sheet 17, Five-in-a-Row).

At the end of Choice Time, be sure to collect the Five-in-a-Row Boards for use again during Choice Time in Sessions 7–9.

## Choice 2: Story Problems

**Materials:** Story Problems, Set A, sorted into 6 envelopes; counters (available); unlined paper; paste or glue sticks

Students select one problem at a time from the envelopes of problems you have prepared. They paste it onto the paper they will use. They may work in pairs or individually. When they have found an answer, they record their work in a way that shows how they found it. If students work individually at times and in pairs at other times, they gain experience both working alone and collaboratively on recording their work.

You will need to set up procedures for getting help in reading the problems. You might designate certain students as reading helpers, or start off a group of students yourself by reading a problem with them.

❖ **Tip for the Linguistically Diverse Classroom** If you have the support of second-language parents or other translators, this activity will go more smoothly. Otherwise, provide a set of story problems with simple drawings added to help make them comprehensible, or act out each problem with concrete materials.

Also establish procedures for how students record the problems they complete. You might ask students to keep a sheet for Set A and record the number of each problem they complete. Alternatively, you might provide a sheet that lists the story problems in this investigation, and students can check off or circle problem numbers as they complete them (see example on page 146).

The **Teacher Note,** About the Story Problems in This Investigation (p. 145), describes the types and level of difficulty of the problems in Set A (as well as Sets B and C, to be used later), and suggests ways to structure student work on them.

Choose two problems of different types, such as problems 2 and 4, for all students to solve. Many students will have time to solve more than two problems. You might guide students in selecting problems at the appropriate level of challenge, or let them make their own choices.

**Note:** If you noticed some students having difficulty with the problems in Sessions 1 and 2, you may want to adjust the numbers in the Story Problems, Set A for those students.

Students continue working on the problems for homework. Some may not finish all six Set A problems by the end of Sessions 3–5; they can continue to work on them during the next Choice Time in Sessions 7–9.

## Observing the Students

Circulate to observe students at work. Be sure to observe student work on the story problems carefully in preparation for the Teacher Checkpoint, Student Strategies (further described on p. 143).

### Five-in-a-Row

■ How do students find sums? Do they count out each quantity? count on from one of the numbers? Do they "just know" some number combinations? Can they quickly add 1 or 2 to another number mentally, or do they need to count it out? Do they use number combinations they know to figure out those they don't know?

- How do students determine which squares to cover? Are they beginning to think strategically about which squares will help them get five in a row?

- Are students able to keep the counters in place? (Students might mark squares by drawing an X to cover a sum, but this will necessitate a new gameboard for each game.)

- Do students play cooperatively and help one another? As needed, remind students that players help one another with the game by thinking through possible moves, explaining their strategies to one another, and waiting while their partners take the time they need to find sums.

**Story Problems**

- Do students understand whether quantities are being combined or separated? Do they know what they need to find in order to solve the problem? Can they keep the situation in mind as they solve the problem?

  As needed, remind students to think through the situation before they solve the problem. How does the problem start? Then what happens? Do you end up with more or less than what you started with? Encourage them to act out the problem with a partner, to build a model of the situation with counters, or to draw pictures of the situation.

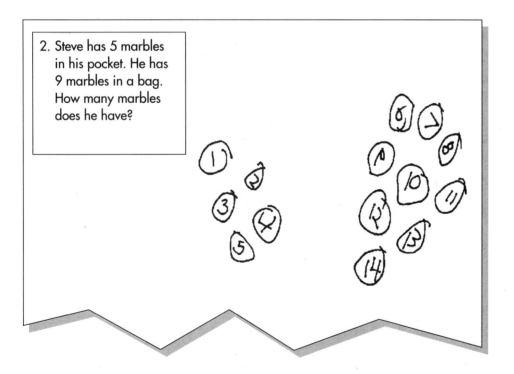

2. Steve has 5 marbles in his pocket. He has 9 marbles in a bag. How many marbles does he have?

- What strategies do students use to solve the problem? Do they model the problem with pictures or counters? Do they count out each quantity? Do they count on or count down from one number? Can they count on or down accurately, knowing where to begin and end? Do students "just know" some of the number combinations? Do they use number combinations they know to figure out those they don't know?

- Are students beginning to take numbers apart in ways that help them solve the problem more easily? (For example, to add 8 + 7, do they break 7 into 2 and 5?) Can they keep track of the parts of numbers they break up and remember what to do with them?

- How clearly can students show their strategies? How are students using numbers or equations as part of their recording?

  Some students may spend time making elaborate pictures of the situation. Ask them how they would use their drawings to find a solution. For some students, making detailed drawings is an important way to make sense of and think through the problem situation. For a few, though, it may be a way to delay the hard work of solving the problem.

- Are the numbers in the problems of average difficulty for most students? Do some students need smaller or larger numbers?

  Most students should start with the numbers given in the problems, but as you watch them at work, you can modify the numbers where appropriate. For more challenge, you might ask some students to find and record more than one way to solve a problem.

## Activity

## Sharing Story Problem Strategies

About halfway through Session 5, call students together to share strategies for the problems you asked all students to complete. Ask for a few volunteers to explain how they solved the problem, and record each method that students suggest on the board or on chart paper.

Encourage students to think about each others' ideas by occasionally asking for volunteers to restate someone else's strategy. For example, after Nathan explains his strategy, you might ask:

**Who can say in your own words how Nathan did this problem?**

Discourage any apparent reliance on individual words in the problems, such as *how many... left,* that might seem to signal a particular operation. The **Teacher Note**, "Key Words": A Misleading Strategy (p. 147), explains why this approach can sometimes interfere with understanding the structure of the whole problem.

To get a sense of how students are solving combining and separating problems, you need to observe students during Choice Time as well as review their written work on the story problems. As you observe, you may want to use a class list to make notes on each student, jotting down which problems they are doing successfully, which cause them difficulty, and what strategies they are using. If possible, observe each student working on both a combining and a separating problem. At the end of Session 5, collect students' work.

Notice which students are counting all, which are counting on, and which are breaking up numbers in flexible ways. For example, suppose you are reviewing students' work on problem 2, a combining problem:

Steve has 5 marbles in his pocket. He has 9 marbles in a bag. How many marbles does he have?

You might see strategies like these:

■ Some students may count out a group of 5 things and another group of 9 things and count them all from 1.

■ Some students may count up 5 from 9, or count up 9 from 5. They might count mentally, or with the help of their fingers or a number line.

■ Some students may reason about number combinations and number relationships. For example: "I know 9 and 1 is 10, and 4 more is 14." Or, "I know 9 is 5 and 4, so 5 and 5 is 10, and 4 more is 14."

This checkpoint can help you think about how the class as a whole is doing and what individual students may need as they continue to work on story problems during Choice Time in Sessions 7–9. For example, if some students are having difficulty counting and keeping track of their counts, you might pull them together in a small group to work through a problem or two with you. You might also adjust the level of challenge in some of the problems or make up your own problems for them. If you have a few students who are comfortable reasoning about number combinations, challenge them with problems that use larger numbers. Students just beginning to use strategies that involve reasoning about number combinations can continue solving problems at about the same level of difficulty, so that they can develop their strategies while working with familiar numbers.

This checkpoint can also help you decide which two or three problems in Set B all students should solve during Sessions 7–9. Choose problems of average difficulty for most of the students in your class.

## Teacher Checkpoint

## Student Strategies

# Sessions 3, 4, and 5 Follow-Up

🏠 **Homework**

**Finishing Story Problems** During these Choice Time sessions, students might take home story problems from Set A (pasted to unlined paper) to complete as homework.

**Five-in-a-Row** Students teach someone at home to play Five-in-a-Row. They will need Student Sheet 17, Five-in-a-Row, one copy each of Student Sheets 18–20, Five-in-a-Row Boards A–C, two number cubes (or the number cards 1–6 from their home sets), and counters (such as buttons or pennies) to use as game markers.

## About the Story Problems in This Investigation

Two sets of story problems are used during Choice Times in this investigation, with an optional third set provided for students who need more challenge. A summary of the types of problems in each set is given below. This information is to help you quickly identify problems of a particular type as you observe students working. Students should not be told what types of problems these are, as they need to think about the situation the problem is describing and then determine how best to solve it.

**Number Size** In students' first experiences with story problems, they need to focus on the often challenging work of making sense of story problem situations, developing solution strategies, and recording their thinking. For this reason, the problems in Sets A and B involve numbers 20 and under. Over time, as students become more comfortable with story problems and develop more knowledge of number combinations and relationships, they will be ready to work with a wider range of numbers.

Students clearly ready to work with larger numbers can solve the problems in Set C when they have completed their work on Sets A and B.

All students will work with a wider range of numbers and problem types in the later grade 1 unit *Number Games and Story Problems.*

### Story Problems, Set A

*Problems 1–3: Combining with an unknown outcome.* Students find the total of two amounts.

1. Rosa had 6 toy cars. Her mom gave her 6 more. How many toy cars does Rosa have now?

2. Steve has 5 marbles in his pocket. He has 9 marbles in a bag. How many marbles does he have?

3. A squirrel ate 8 nuts. Then she ate 7 more nuts. How many nuts did she eat?

*Problems 4–6: Separating with an unknown outcome.* Students find how many are left when one amount is removed from another.

4. Rosa had 11 library books. She took 4 of them back. Now how many books does she have?

5. Steve and Rosa had 12 apples. They ate 6 of them. How many apples are left?

6. Rosa had 15 pennies. She spent 6 pennies. How many pennies did she have then?

### Story Problems, Set B

*Problems 1–2: Combining with an unknown outcome.*

1. 10 children were playing at the park. Then 8 more came to play. How many children were at the park?

2. Kim and Tito picked 11 red flowers. They picked 5 white flowers. How many flowers did they pick?

*Problem 3: Combining three amounts with an unknown outcome.*

3. Kim has 5 cookies. Tito has 4 cookies. Jill has 3 cookies. How many do they have in all?

*Problems 4–6: Separating with an unknown outcome.*

4. Kim had 16 pennies in her pocket. The pocket had a hole. 8 pennies fell out. Now how many pennies are in her pocket?

5. Tito had 14 fish in a tank. He gave away 9 fish. How many fish does he have now?

6. Kim saw 20 ducks on the pond. Then 9 ducks flew away. How many ducks were still on the pond?

*Continued on next page*

### Story Problems, Set C (Challenges)

Reserve these problems for students who complete all of Set A and Set B and who need more challenge.

*Problem 1: Combining with an unknown outcome.*

1. Steve and Rosa picked cherries. They ate 18 cherries for lunch. Rosa ate 6 cherries after lunch. How many cherries did they eat?

2. Rosa had 11 stamps. Her father gave her 12 more. Now how many stamps does Rosa have?

*Problem 3: Combining three amounts with an unknown outcome.*

3. 4 children were on the bus. At the next stop, 13 more got on. Then 6 more children got on. How many children were on the bus?

*Problems 4–6: Separating with an unknown outcome.*

4. Steve had 19 pennies. He spent 15 pennies. Now how many does he have?

5. Rosa had 21 balloons. She gave 3 of them away. Now how many does she have?

6. Rosa's mom had 23 hens. She sold 8 of them. Then how many hens did she have?

**Guiding Student Work** For each Choice Time, you should specify two or three problems for all students to attempt; these problems are the focus of a whole-class discussion at the end of that Choice Time. Otherwise, students may choose freely among the other problems (including any not completed from the previous Choice Time). When students are allowed to choose from a variety of problems, they learn to find their own level of challenge.

To help students keep track of which problems they have done, you might provide a simple form.

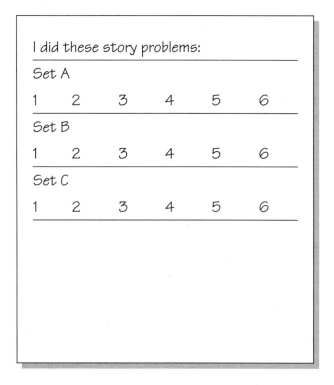

There may be times when you want to assign particular problems to students who will not have time to complete all the problems in this investigation, and who could use more work with certain problem types or with numbers of a certain size.

To help students who are having difficulty, you might call together small groups to work with you on a problem during Choice Time; you might adjust the numbers in some problems; or you might create some problems of your own. Remember that while it's fine to alter the numbers or the context of the problems, even a slight change in the problem structure can mean a great change in the level of difficulty. For more information, review the **Teacher Note**, Types of Story Problems, p. 123.

# "Key Words": A Misleading Strategy

Some mathematics materials have advocated a "key words" technique to help students solve story problems. Students are taught to recognize words in a problem that provide clues about how to choose which operation to use to solve it. For example, *altogether* or *more* signal addition, whereas *left* or *fewer* signal subtraction.

> I had 5 marbles. Lee gave me 6 more. How many do I have *altogether?*

> I had 16 marbles. I gave away 8. How many do I have *left?*

> I had 16 marbles. Lee has 7 *fewer* than I do. How many marbles does Lee have?

There are two flaws in the key word approach. First, these words may be used in many ways. They might be part of a problem that requires a different operation from the expected one:

> There are 28 students in our class altogether. There are 13 boys. How many are girls?

If we trust in key words, then *altogether* in this problem should signal addition of the numbers in the problem, 28 + 13. In fact, the problem calls for finding the *difference* between 28 and 13.

The second reason for avoiding reliance on key words is that students should think through the entire structure of the problem. They need to read the problem and understand the situation so that they can construct a model of the problem for themselves. Here's another example:

> I am making cookies for my party. There will be 6 people at my party, including me. I want each person to have 4 cookies. How many cookies should I make altogether?

If students are encouraged to use key words, they are likely to simply pull numbers out of the problem and carry out some operation (in this case, perhaps, 6 + 4) without developing a model of the whole problem (in this case, 6 equal groups of 4).

# Dot Addition

## Materials

- Transparencies of Dot Addition Cards
- Overhead projector
- Dot Addition Cards (1 per pair, plus 1 per student, homework)
- Student Sheet 21 (1 per student, homework)
- Student Sheet 22 (1 per pair and a few extras for class, plus 1 per student, homework)
- Lined or unlined paper
- Blank Dot Addition Board (several per class, optional)

## What Happens

Students play Dot Addition, in which they combine dot cards to make given numbers. Their work focuses on:

- finding combinations of numbers up to about 15
- finding the total of several small, single-digit numbers
- using dot patterns to model number combinations

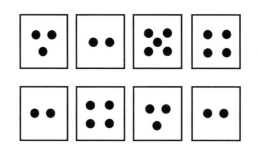

## Activity

### Introducing Dot Addition

Introduce the Dot Addition Cards, which are similar to the Dot Cards students used in Investigation 1, by briefly doing Quick Images (see p. 6) with the transparent set. Then explain to the class that today, they'll be learning a new activity with dot patterns.

Bring a deck of Dot Addition Cards and Dot Addition Board A to the meeting area. Lay out three 2-dot cards, two 3's, two 4's, and one 5 on the floor (as shown above). Mix the cards up, rather than grouping them by type.

**Who sees a way to put together some cards to make 6? … Leah says 2 and 2 and 2. Can you show us how you know 2, 2, and 2 make 6?**

**So, Leah says that she knows 2 and 2 is 4, and then she counted up one more 2: 5, 6. Does anyone have another way to prove that 2 and 2 and 2 makes 6? Does anyone see a different way to use the cards to make 6?**

Encourage the class to share two or three different ways to show that each combination is correct.

After students have suggested a few combinations of 6, ask them to find ways to put the cards together to make 8.

**Demonstrating the Dot Addition Game**  Set up a sample game. Explain that each pair gets a set of dot cards with 2–5 dots. At the start of the game, pairs lay out these cards in rows of five. Begin laying out a set of dot cards faceup. After you have put out two rows of cards and part of a third, pause and ask students to figure out how many cards are out. Ask for a few volunteers to share their thinking. Do they count each card? Do they count by 5's? Do they know the number combination 5 + 5?

When you have put out four rows of five cards, ask students to figure out how many cards are out and take a few moments for some students to share strategies.

Now lay out Dot Addition Board A next to the dot cards. Explain that the object is to move cards onto the board to make a combination for each number on the board. Players can't use any card twice. However, they can rearrange their cards at any time until they have filled the board.

As you demonstrate the game, involve students by asking for a volunteer to suggest one combination for each of the numbers 6, 8, 10, and 12.

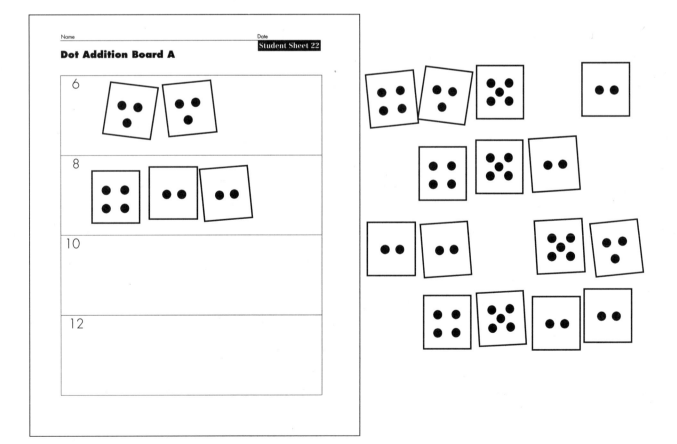

When you have finished the board, show them how to record each combination with addition notation on a separate sheet.

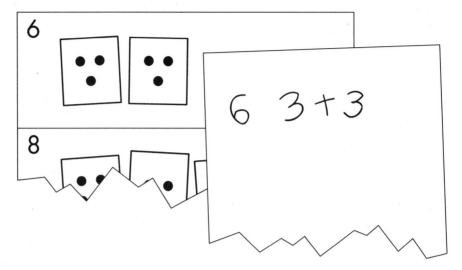

---

## Playing Dot Addition

Distribute Dot Addition Board A and the Dot Addition Cards to each pair of students. Pairs lay out their cards faceup in four rows of five. They find combinations of dot cards for each number on the board. When they finish, each student records the addition combinations on another sheet.

If some students find the numbers on Dot Addition Board A very difficult, use the Blank Dot Addition Board to make boards with smaller numbers, such as 5, 6, 7, and 8, or 6, 6, 8, and 8.

**Note:** If students want to make combinations that involve more than four numbers, such as 2 + 2 + 2 + 2 + 2 for 10, they will have to either overlap their Dot Addition Cards or continue placing them in a row that goes off the board. Assure them that this is fine.

### Observing the Students

As students play the Dot Addition game, watch for the following:

- How do students construct their sums? Do they take a random approach, trying some cards, then discarding those and trying different cards, until they find a combination that works? Do they begin with one number, then count up to see what other number they need? Do they look for particular combinations? ("I need to make 6. Do I have a 2 and a 4?")

- When counting to find a total, do they always count out each quantity from 1? Do they begin with the number of dots on one card and count up? Do any students count by numbers other than 1?

Some students may begin counting by 2's or 3's for small quantities, and then switch to 1's as quantities increase. Understanding of counting by numbers other than 1 develops gradually over the elementary years. Students will have opportunities to develop and extend this understanding in the grade 1 unit *Number Games and Story Problems* as well as in grades 2 and 3 of the *Investigations* curriculum.

■ Do students use knowledge of number combinations? What number combinations do they know? Do they know doubles (such as 3 + 3 or 4 + 4)? Can they quickly add 1 or 2 to another number?

■ Are they comfortable recording with addition notation? You might need to remind students to write the total next to each combination. If the notation confuses some students, they might leave out the plus (+) sign and just record the numbers they used to make up the totals.

If you can't tell how students are making their sums, ask them to explain their strategies. Ask, for example:

**How did you figure out this combination for 6?**

**What if you couldn't use any 5's. Could you figure out something else that would work for 10?**

**I see you've started with a 5 card to make 8. How are you figuring out what else you need?**

If any students finish in this session, ask them to play again without using *any* of the combinations they used in their first game. That is, if they used three 2's to make 6 the first time, they need to find a new way to make 6 this time. The same rules apply: They place the cards on the board, and they can't use a card more than once.

Be sure students save Dot Addition Board A for use again during Choice Time in Sessions 7–9.

## Session 6 Follow-Up

**Dot Addition** Students teach someone at home to play Dot Addition. They will need to take home Student Sheet 21, Dot Addition (the directions); Student Sheet 22, Dot Addition Board A, and Dot Addition Cards. If you think students won't have scissors at home to cut apart the cards, find a time when they can do this at school.

 **Homework**

# Combining and Separating

## What Happens

Students do Quick Images with combinations of dot cards. They add Dot Addition to their other Choice Time activities, Solving Story Problems and Five-in-a-Row. At the end of Session 8, students share strategies for finding addition combinations in the games. At the end of Choice Time, they again share strategies for solving story problems. Their work focuses on:

- becoming familiar with single-digit addition pairs
- developing strategies for solving combining and separating story problems
- recording strategies for solving combining and separating story problems using pictures, numbers, words, and equations
- finding combinations of numbers up to about 15
- finding the total of several small, single-digit numbers
- using dot patterns to model number combinations

## Materials

- Transparencies of Dot Addition Cards
- Five-in-a-Row Boards A–C (from Sessions 3–5)
- Number cubes (12–16 for the class)
- Story Problems, Sets A–C (in prepared envelopes)
- Dot Addition Board A (from Session 6)
- Student Sheets 23–24 (1 each per pair and a few extras for class, plus 1 each per student, homework)
- Blank Dot Addition Board (several for the class, optional)
- Dot Addition Cards (1 per pair)
- Paste or glue sticks
- Paper, lined and unlined

## Activity

### Dot Addition Quick Images

Present three or four Quick Images to the class. Each time, use combinations of two or three of the same cards, with up to 5 dots. For example:

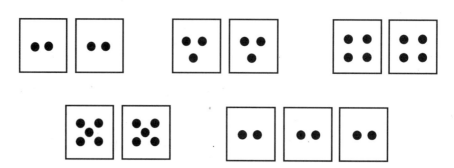

If you have time for Quick Images again later in the day, outside of math time, continue with some of these more challenging combinations:

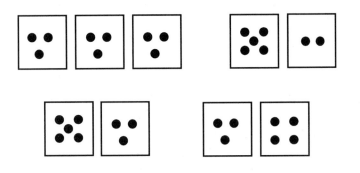

## Choice Time

Most of Sessions 7–9 will be spent in Choice Time. Add a new third choice, Dot Addition, to those already posted.

Plan to gather students near the end of Session 8 to share strategies for finding combinations as they play Dot Addition. Announce the end of Choice Time midway through Session 9, and spend the rest of the session with students sharing strategies for two or three of the story problems they solved.

To review Choice 1, Five-in-a-Row, see p. 139.

## Choice 2: Story Problems

**Materials:** Story Problems, Set B, sorted into six envelopes, and any remaining problems from Set A; counters; unlined paper; paste or glue sticks. (Have Story Problems, Set C, available for any students ready for more challenge.)

Students continue solving and recording solutions to story problems. Choose two or three of the problems in Set B for all students to attempt, including a *combining with unknown outcome* problem (such as problem 1) and a *separating with unknown outcome* problem (such as problem 4). You might also ask students to work on problem 3, in which they combine three amounts to find the total. At the end of Session 9, you will be calling students together to share their strategies for solving these problems.

Most students will have time to work on several problems. You might guide them in selecting problems at the appropriate level of challenge, or let them make their own choices. Students may continue working on the problems for homework. Some students may not have time to complete all the story problems in Sets A and B.

Students who finish early and are ready for more challenge can work on the Story Problems, Set C. Other students who finish early could find more than one way to solve some of the problems in Sets A and B. Or, make up more problems they can solve, using numbers about the same size as those in Sets A and B.

## Choice 3: Dot Addition

**Materials:** Dot Addition Cards (1 set per pair); Student Sheets 22–24, Dot Addition Boards A–C (1 each per pair); lined or unlined paper

Pairs lay out their cards faceup in four rows of five, and place one of the Dot Addition Boards beside them. They find combinations of dots for each number on the board, using each card only once. When they finish, each student records the addition combinations on another sheet.

Students choose from among the three boards. By the end of Choice Time, they should have worked with at least one board besides Board A. If they have time to use some boards twice, the second time they should find different combinations for each number.

**Different Levels of Challenge** You can use copies of the Blank Dot Addition Board to adjust the level of difficulty of the game. For more practice, make boards with smaller numbers (5, 6, 7, and 8, or 6, 6, 8, and 8). For more challenge, use larger numbers (12, 15, 16, and 18, or 12, 12, 16, and 16).

Another variation that offers more challenge is to try making the numbers on the board with all the same cards (for example, all 2's or all 3's). On Board A, for example, they can make the 6 with two 3's, the 8 with two 4's, the 10 with two 5's, and the 12 with six 2's. (Note that it is not possible to make all the numbers on Board C with the same number unless you combine decks of Dot Addition Cards.)

## Observing the Students

For a review of questions to guide your observations for the activities, refer to the following: Choice 1: Five-in-a-Row (p. 140); Choice 2: Story Problems (p. 141); Choice 3: Dot Addition (p. 150).

Activity

# Strategies for Finding Combinations

About 15–20 minutes before the end of Session 8, gather students for a brief discussion of their strategies for finding combinations when playing Dot Addition. Students need not have completed their work on this choice in order to participate in the discussion.

Begin by asking for volunteers to tell one way to make 6 with the dot cards. Students may refer to the combinations they recorded, or they may use dot cards, fingers, or other strategies. Encourage students to find several ways to show that each combination is correct.

**Yanni says he can make 6 with 2 and 4. How do you know that makes 6?... OK, so Yanni knew that 2 and 3 is 5, and then he counted up 1 more. Who has another way they know 2 and 4 is 6?... Chanthou held up 2 fingers on one hand, and 4 on the other like this, and then counted all the fingers. Who has another way?**

Gather two or three ways to show that each combination is correct. If students suggest using combinations they know to find new ones, point out that building from combinations you know is a good way to find totals. However, keep in mind that some of your students may not yet reason in this way. For them, each combination is a new problem to be solved by counting all the dots (or counters or fingers). Students will begin recognizing relationships among combinations as they continue through the early elementary grades. See the **Teacher Note**, Strategies for Learning Addition Combinations (p. 158), for more information on learning addition combinations (number combinations with two addends) at this level.

Record each combination that students suggest. After you have recorded a few combinations, if no one has suggested a way to make 6 with all of one number (all 2's or all 3's), propose this idea:

**Did anyone make six with all of one number?... Diego made six with three 2's. How can you prove that three 2's is 6?**

**OK, if you know 2 and 2 is 4, you can count up two more, 5, 6. Who has another way?... So, Eva counted by 2's: one, TWO, three, FOUR, five, SIX** *[emphasizing alternate numbers].* **Let's count by 2's like Eva did.**

Remember that counting by 2's (or another number) may not yet be meaningful for all students. Some may know the first few numbers in the counting-by-2's sequence, but may not yet connect this with the process of adding 2 more with each number they say. Students will continue to develop meaning for counting by numbers other than 1 over the next couple of years.

Repeat the discussion for the number 7, which does not appear on any of the three boards. Encourage students to work mentally or to use dot cards or fingers to find combinations of 7. Emphasize any strategies that involve building from combinations of 6. For example, students might note that they can make 7 with 3 and 4, because they know 3 and 3 is 6, and 7 is one more.

If time permits, repeat the activity again for 8. As you did with 6, encourage students to find ways to make 8 using all of just one number.

To challenge students, ask for their ideas about what other numbers they can make with all 2's or all 3's:

**So, we can make 6 with three 2's and 8 with four 2's. What number can we make with five 2's? What are some other numbers we could make with just 2's?**

## Activity

## Sharing Story Problem Strategies

About halfway through Session 9, end Choice Time and gather students to share strategies for the problems you asked everyone to complete. Ask for a few volunteers to explain how they solved the problem. Record each method that students suggest on the chalkboard or on chart paper. Whenever possible, base your recording on the way the student recorded.

**Set B, problem 3:** Kim has 5 cookies. Tito has 4 cookies. Jill has 3 cookies. How many do they have in all?

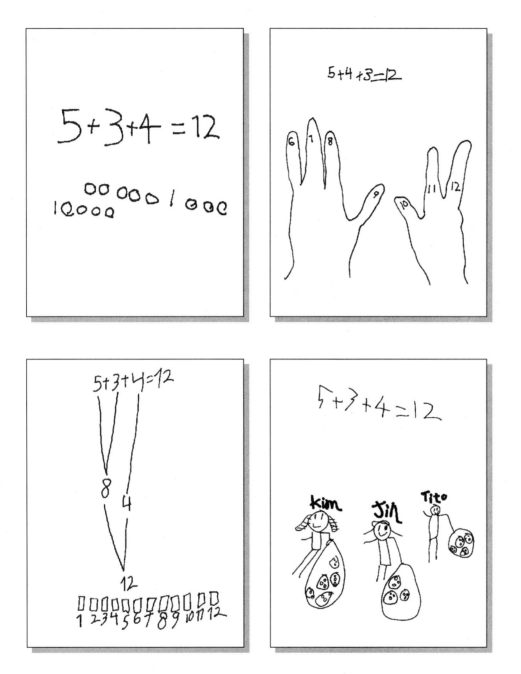

# Sessions 7, 8, and 9 Follow-Up

**Homework**

**Story Problems** Students take home story problems from Set A and Set B that they started in class. They should be sure to bring these back to keep in their folders with other story problems they have solved, and to record these problems as completed.

**Dot Addition** Send home Student Sheets 23 and 24, Dot Addition Boards B and C, for another game of Dot Addition at home. They will also need Dot Addition Cards, which they may have already taken home. You might ask students to return the sheets on which they record the combinations they make for each game.

**Extension**

**What's a Story for This Problem?** Write an addition expression, such as 3 + 9, on the board or on chart paper. Ask students to make up their own story problem for the expression. They may use pictures, words, or both to record the problem. They also solve their problem and show how they solved it.

---

**Teacher Note**

## Strategies for Learning Addition Combinations

In order to develop good computation strategies, students eventually need to become fluent with the addition combinations (number combinations with two addends, also called addition pairs) from 0 + 0 to 10 + 10. These combinations are part of the repertoire of number knowledge that contributes to the rich interconnections among numbers that we call *number sense*. A great deal of emphasis has been put on learning these combinations in elementary school. While we agree that knowing these combinations is important, we want to stress two important ideas:

■ Students learn these combinations best by using strategies, not simply by rote memorization. Relying on memory alone is not sufficient, as many of us know from our own schooling. If you forget, as we all do at times, you are left with nothing. If, on the other hand, your learning is based on your understanding of numbers and their relationships,

you have a way to rethink and restructure your knowledge when you don't remember something you thought you "knew."

■ Knowing the combinations should be judged by fluency in use, not necessarily by instantaneous recall. Through repeated use and familiarity, students will eventually come to know most of the addition combinations immediately, and a few by using some quick and comfortable numerical reasoning strategy. For example, when one of the *Investigations* authors thinks of 8 + 5, she doesn't automatically see the total as 13; rather, she sees the 5 broken apart into a 2 and a 3, the 2 combined with the 8 to make 10, then the 10 and the 3 combined to total 13. While this strategy takes quite a while to write down or to read, she "sees" this relationship almost instantaneously. As far as she is concerned, she "knows" this combination.

We avoid calling these addition combinations the addition "facts," because we think the term "facts" tends to elevate knowledge of these combinations above other mathematical knowledge, as if knowing these is the most important thing in mathematics. Developing fluency in using these combinations is important, but many other ideas are just as critical for building number sense.

## Using Strategies to Learn Addition Combinations

At grade 1, many students are still primarily figuring out number combinations by counting by 1's. They are not yet thinking about larger units, so they are not yet ready to use strategies to learn number combinations. However, if you see some students begin to develop strategies based on number relationships, you can encourage them to articulate and build on this thinking. Students at this age begin to use several important strategies:

■ **Adding 1's and 2's** Adding 1 or 2 easily is a beginning step in learning addition combinations. Students begin to "just know" that any time you add 1 to a number, you end up with the next number in the counting sequence. After students have had many experiences adding on 1 object to a group of objects, they are able to coordinate the idea of adding on 1 with their knowledge of the counting numbers. Building on this idea, they begin to add on 2 by quickly counting on two more numbers.

■ **The Doubles** Students learn many of the doubles (3 + 3, 4 + 4, 5 + 5, and so forth) quite early. Some students can begin to use the doubles they know to help them figure out other combinations: "I know that 5 + 5 is 10, so 6 + 5 is one more, that's 11."

■ **Sums That Make 10** Some students may begin to recognize some of the combinations that make 10 and can eventually build on these to find others. For example, knowing that 6 + 4 is 10, students may be able to imagine taking 1 from the 4 and adding it to the 6, giving them 7 + 3.

Encourage your students to share their strategies for learning number combinations and to explain how they figured out new or difficult combinations. Keep in mind, though, that first grade students are just becoming familiar with number combinations. By the end of first grade, most will be comfortable with combinations that involve +1 and +2, and many will also know quite a few other combinations, including several doubles and combinations of 10. But many students will not yet be able to use the combinations they know to find other combinations, and most will not yet recognize that, for example, adding 2 + 8 is the same as adding 8 + 2. Your students will have many opportunities to continue work on number combinations in second grade and the early part of third grade.

# Solving Story Problems

## Materials

- Story Problems, Set D, cut apart
- Unlined paper
- Paste or glue sticks
- Hundred Number Wall Chart with colored chart markers

## What Happens

As an assessment, students solve story problems and record their solution strategies. For the remainder of the session, they do Clapping Patterns. Their work focuses on:

- solving combining and separating story problems
- recording strategies for solving combining and separating story problems using pictures, numbers, words, and equations

## Activity

### Assessment

### Solving Story Problems

Distribute Story Problems, Set D (for Assessment) one problem at a time. As before, students paste each problem on a sheet of paper, solve it, and record their solution strategies. While you don't want to prevent students from talking to each other about the problems since that would establish an unusual atmosphere, tell them that you want each of them to find their own way to solve each problem and to record their strategies clearly so that you can see how they are thinking about the problems.

Ask all students to solve problem 1, a combining problem with an unknown outcome, and problem 2, a separating problem with unknown outcome. Any who finish early may also solve problem 3, a combining problem in which three amounts are combined with an unknown outcome.

If you feel that the problems are too difficult for some students, adjust the numbers or substitute your own problems. Students may use counters to help them.

You may also decide to use larger numbers for some students. However, students may rely on counting by 1's for problems involving larger or less familiar numbers. If they work with smaller numbers, they may be more likely to use strategies that involve knowledge of number combinations and number relationships.

### Observing the Students

Consider the following questions as you observe students working. See also the **Teacher Note**, Assessment: Solving Story Problems (p. 165), for suggestions for assessing students' work.

- Are students comfortable with both combining and separating problems? Can they make sense of what happens in each problem? Do they know what they need to find in order to solve the problem? Can they keep the situation in mind as they solve it?

- What strategies do students rely on? Do they count out all the quantities in the problem? Do they start with one quantity and count on or down? Do they use number combinations they know to solve the problem? Are students taking numbers apart in ways that help them solve problems more easily? Can they keep track of the parts of numbers they create and what to do with them?

- Can students record their strategies in a way that makes sense, using some combination of pictures, words, and numbers?

The following examples of student work show a variety of strategies.

**Set D, problem 1:** Kim made 7 paper hats. Tito made 6 paper hats. How many paper hats did they make?

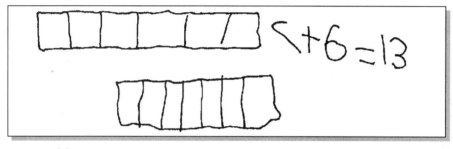

$7+6=13$

Direct modeling

I stated at 7 and I counted up 6 and I got to 13

Counting

I know thet $6+6=12$ if I add 1 more It is 13

Numerical reasoning

**Set D, problem 2:** Tito had 15 cookies. He gave 9 to his friends. How many cookies did Tito have then?

Direct modeling

Counting

Counting: "I used the number line"

Numerical reasoning

# Clapping Patterns

For the remainder of the session, do Clapping Patterns with your class. (As needed, review the directions on p. 106.)

Choose two or three patterns based on four beats and two or three actions; for example: knees-knees-knees-clap, clap-clap-knees-knees, and knees-knees-knees-tap head. Act out the patterns with the class, and then ask students how the patterns are alike and different. Begin by asking them to compare just two of the patterns:

**How are knees-knees-knees-clap and clap-clap-knees-knees alike? How are they different?**

**Kristi Ann says both have four parts. OK, who notices something else?**

**Yanni says both use knees and claps. Anyone notice anything else?... Jonah says one has one clap and the other has two claps.**

Your students will probably notice that both patterns have four beats and that they use different numbers of each action to make up four: knees-knees-knees-clap uses three of one action and one of another to make up four beats; clap-clap-knees-knees uses two of each to make up four beats.

If your students don't relate the patterns to combinations of four, encourage them to do so:

**Jonah noticed that one pattern has one clap and the other has two claps. What about knees? Do they have the same number of knees?**

**So, Max noticed that knees-knees-knees-clap has three knees and one clap, and clap-clap-knees-knees has two knees and two claps.**

Ask students to think of other clapping patterns with four beats. Act out each clapping pattern with the class a few times, and then discuss similarities and differences among them. Again, encourage students to relate the patterns to combinations of four.

In any remaining time, choose one of the patterns and work with the class to represent it with cubes and with the 100 chart, as you did in Investigation 3.

## Choosing Student Work to Save

As the unit ends, you may want to use one of the following options for creating a record of students' work on this unit.

- Students look back through their folders or notebooks and think about what they learned in this unit, what they remember most, what was hard or easy for them. Students might discuss this with a partner or share in the whole group.

- Depending on how you organize and collect student work, students might select some examples of their work to keep in a math portfolio. In addition you may choose some examples from each student's folder to include. Their work on Twelve Cats and Dogs (p. 38), Three Towers (p. 75), and Solving Story Problems (p. 160) can be useful pieces for assessing student growth over the school year. You may want to keep the original and make copies of these pieces for students to take home (or vice versa).

- Send a selection of work home for families to see. Some teachers include a short letter, summarizing the work in this unit. You could enlist the help of your students and together generate a letter that describes the mathematics they were involved in. This work should be returned to you if you are keeping a year-long portfolio of mathematics work for each student.

# Assessment: Solving Story Problems

At grade 1 it is difficult to assess students solely from their written work, as there is a gap between their thinking and what they are able to put down on paper. Make a point to observe students working on the story problems for the final assessment activity and jot down notes about how each student is approaching the task (use the questions on page 161 as guidelines). A combination of your observations and students' written work can give you a sense of students' strategies for solving combining and separating problems.

You may find that students fall into three general groups. In one group are students who are doing competent and appropriate work. These students can figure out what they need to do to solve the problem, can select an appropriate strategy, and can record their work. They may "just know" some number combinations, but they tend to rely primarily on strategies that involve counting by 1's. For example, they may combine 6 and 7 by drawing a group of 6 things and a group of 7 things and then counting everything. Or, they may start at 7 and count up 6, perhaps using their fingers to help them keep track of the count. They may take 9 from 15 by counting out 15 objects, removing 9, and counting the remaining set. Or, they may start at 15 and count down 9, perhaps using a list of numbers they have written or a number line to help them.

In another group are students who, in addition to what the first group can do, are using good numerical reasoning. They might combine 6 and 7 by breaking 7 into 4 and 3, because they "just know" that 6 and 4 is 10, and then adding on 3 more. Or, they might explain that they know that 6 and 6 is 12, and 1 more is 13. They might take 9 from 15 by breaking 9 into 5 and 4, and then explaining that they know that 15 take away 5 is 10, and 4 less than 10 is 6.

In a third group will be students about whom you have some concerns. They may have some difficulty understanding how to approach problems, especially those that involve separating, and may have trouble explaining and recording their strategies. These students may not be comfortable working with numbers in the teens or 20's. They may have trouble keeping track of their counts in this range, especially when counting backwards, and they may not be able to count out 10 or more objects accurately.

These three groupings can help you think about the class as a whole and about what individual students may need when your work on story problems continues in a later unit, *Number Games and Story Problems*. This overview of the class might suggest further experiences that would be helpful to students over the coming months. For example, some students may benefit from more practice with games that involve counting and combining, such as Double Compare (or the easier version, Compare), Five-in-a-Row, or Dot Addition with smaller numbers (such as 5, 6, 7, and 8). Other students may be ready to tackle more challenging variations of these activities, such as Triple Compare, the more difficult three-number-cube variation of Five-in-a-Row, and Dot Addition with the larger numbers (such as 13, 15, 16, and 18).

Much of the work presented in this unit is followed up and extended in the *Investigations* grade 1 unit *Number Games and Story Problems*. If you are following the suggested grade 1 sequence, that unit is taught during the second half of the school year. Students will, however, benefit from having consistent opportunities to work on the combining and separating strategies they have begun to develop in this unit.

Throughout the school year, present students with combining and separating situations in the form of written problems. Also look for opportunities in the classroom where adding or subtracting numbers naturally occurs.

# Teacher Note ⟩ *About Choice Time*

Choice Time is an opportunity for students to work on a variety of activities that focus on similar mathematical content. Choice Times are found in every unit of the grade 1 *Investigations* curriculum. These generally alternate with whole-class activities in which students work individually or in pairs on one or two problems. Each format offers different learning experiences; both are important for students.

In Choice Time the activities are not sequential; as students move among them, they continually revisit some of the important concepts and ideas they are learning. Many Choice Time activities are designed with the intent that students will work on them more than once. As they play a game a second or third time, or as they work to solve similar problems, students are able to refine their strategies, see a variety of approaches, and bring new knowledge to familiar experiences.

You may want to limit the number of students working on a particular Choice Time activity at any one time. In many cases, the quantity of materials available limits the number. Even if this is not the case, limiting the number is advisable because it gives students the opportunity to work in smaller groups. It also gives them a chance to do some choices more than one time. Often when a new choice is introduced, many students want to do it first. Assure them that, even with your limits, they will have the chance to try each choice.

Initially you may need to help students plan what they do. Rather than organizing them into groups and circulating the groups every 15 minutes, support students in making their own decisions. Making choices, planning their time, and taking responsibility for their own learning are important aspects of a student's school experience. If some students return to the same activity over and over again without trying other choices, suggest that they make a different first choice and then do the favorite activity as a second choice.

## How to Set Up Choices

Some teachers prefer to have the choices set up at centers or stations around the room. At each center students will find the materials needed to complete the activity. Other teachers prefer to have materials stored in a central location, with students taking the materials to their own desks or tables. In either case, materials should be readily accessible, and students should be expected to take responsibility for cleaning up and returning materials to their appropriate storage locations. Giving a "5 minutes until cleanup" warning before the end of any session allows students to finish what they are working on and prepare for the upcoming transition.

Decide which arrangement to use in your classroom. You may need to experiment with a few different structures before finding the setup that works best for you and your students.

## The Role of the Student

Establish clear guidelines when you introduce Choice Time. Discuss students' responsibilities:

- Try every choice at least once.
- Work with a partner or alone. (Some activities require that students work in pairs, while others can be done either alone or with a partner.)
- Keep track, on paper, of the choices you have worked on.
- Keep all your work in your math folder.
- Ask questions of other students when you don't understand or feel stuck. (Some teachers establish the rule, "Ask two other students before me," requiring students to check with two peers before coming to the teacher for help.)

For each Choice Time, list the activity choices on a chart, the board, or the overhead. Sketch a picture with each choice for students who may have difficulty reading the activity names. Some teachers laminate a piece of tagboard to create a

Choices board that they can easily update as new choices are added from session to session and old choices are no longer offered.

First grade students can keep track of the choices they have completed in one of these ways:

- When they have completed an activity, students record its name or picture on a blank sheet of paper.

- Post a sheet of lined paper at each station, or a sheet for each choice at the front of the room. At the top of each sheet, put the name of one activity and the corresponding picture. When students have completed an activity, they print their name on the corresponding sheet. Keep these lists throughout an investigation, as the same choices may be offered several times.

Some teachers keep a date stamp at each Choice Time station or at the front of the room, making it easy for students to record the date as well.

In any classroom there will be a range of how much work students complete. Some choices include extensions and additional problems for students who have completed their required work. Encourage students to return to choices they have done before, do another problem or two from the choice, or play a game again. You may also want to make the choices available at other times during the day.

Whenever students do any work on paper during Choice Time, they put this in their math folders at the end of the session.

At the end of a Choice Time session, spend a few minutes discussing with students what went smoothly, what sorts of issues arose and how they were resolved, and what students enjoyed or found difficult about Choice Time. Having students share the work they have been doing often sparks interest in an activity. Some days, you might ask two or three volunteers to talk about

their work. On other days, you might pose a question that someone asked you during Choice Time, so that other students might respond to it. Encourage students to be involved in the process of finding solutions to problems that come up in the classroom. In doing so, they take some responsibility for their own behavior and become involved with establishing classroom policies.

## The Role of the Teacher

Choice Time provides you with the opportunity to observe and listen to students while they work. At times, you may want to meet with individual students, pairs, or small groups who need help. This gives you the chance to focus on students you haven't had a chance to observe before, or to do individual assessments. Recording your observations of students will help keep you aware of how they are interacting with materials and solving problems. The **Teacher Note,** Keeping Track of Students' Work (p. 168), offers some strategies for recording and using your observations.

During the initial weeks of Choice Time, much of your time will probably be spent in classroom management, circulating around the room, helping students get settled into activities, and monitoring the process of moving from one choice to another. Once routines are familiar and well established, students will become more independent and responsible for their work during Choice Time. This will allow you to spend more concentrated periods of time observing the class as a whole or working with individuals and small groups.

# Teacher Note  ·  *Keeping Track of Students' Work*

Throughout the *Investigations* curriculum, there are numerous opportunities to observe students as they work. Teacher observations are an important part of ongoing assessment. A single observation is like a snapshot of a student's experience with a particular activity, but when considered over time, a collection of these snapshots provides an informative and detailed picture of a student. Such observations can be useful in documenting and assessing student's growth, as well as in planning curriculum. They offer important sources of information when preparing for parent conferences or writing student reports.

The way you observe students will vary throughout the year. At times you may be interested in particular problem-solving strategies that students are developing. Other times, you might want to observe how students use or do not use materials for solving problems. You may want to focus on how students interact when working in pairs or groups. Or you may be interested in noting the strategy that a student uses when playing a game during Choice Time. Class discussions also provide many opportunities to take note of student ideas and thinking.

You will probably need some sort of system to record and keep track of your observations. While a few ideas and suggestions are offered here, it's important to find a record-keeping system that works for you. All too often, keeping observation notes on a class of 28–32 students can quickly become overwhelming and time-consuming.

A class list of names is one convenient way of jotting down your observations. Since the space is somewhat limited, it is not possible to write lengthy notes; however, over time, these short observations provide important information.

Another common approach is to keep a supply of adhesive address labels on clipboards around the room. After taking notes on individual students, you can peel off each label and stick it in the appropriate student's file.

Some teachers keep a loose-leaf notebook with a page for each student. When something about a student's thinking strikes them as important, they jot down brief notes and the date.

You may find that writing notes at the end of each week works well for you. Some teachers find this a useful way of reflecting on individual students, on the curriculum, and on the class as a whole. Planning for the next week's activities often grows out of these weekly reflections.

In addition to your own notes, you will have each student's folder of work for the unit. This documentation of their experiences can help you keep track of your students, assess their growth over time, and communicate this information to others. An activity at the end of each unit, Choosing Student Work to Save, suggests particular pieces of work you might keep in a portfolio of work for the year.

# First Graders: A Wide Range of Understanding

Students enter first grade with a very wide range of mathematical understanding and experience. At this age, they are in the midst of an important cognitive transition, from reliance on concrete modeling of mathematical situations to the beginnings of reasoning about numbers, quantities, shapes, and other mathematical objects. This transition lasts for several years as students slowly explore how to make mental models of important mathematical relationships. The 4- or 5-year-old who feels confident that "the numbers go on forever," even though she can't actually count past 29 without help, has developed a model of the number system that no longer depends only on what she can see and touch in the world. She is beginning to build an abstract idea about what numbers are and what they do. The 6-year-old who says, "I know that 8 and 5 is bigger than 7 and 5 because 8 is bigger than 7" is reasoning about number relationships. He knows something about the operation of addition and how it works; he realizes that even without figuring out the two sums, he can tell how the two sums relate to each other.

Understandings like these come slowly. Students need many experiences with counting, comparing, building numbers out of cubes or tiles, solving problems, drawing pictures, and talking about their strategies as they develop ideas about the number system and about operations with whole numbers. Some students entering first grade will not be able to reliably count small quantities. Others can easily count or recognize small quantities without counting, but will have more trouble counting quantities greater than 12 or 13. Some students are confident counters, but will rely completely on counting by 1's to solve any problem. Other students are beginning to reason about numbers and operations: "I know that 6 plus 1 is 7, because 7 is the number that comes after 6." And: "I know 6 take away 3 is 3 because 3 plus 3 is 6."

Because students' about number can develop at very different paces throughout first grade and well into second grade, you will need to pay a great deal of attention to individual needs and modify many of the first grade activities so each student is working on the mathematical ideas most critical for him or her. You will want to make sure students have enough challenge but are not overwhelmed. In some activities, we offer particular suggestions for modifying it for individual students working at different levels. Many of the games and activities in this unit and throughout the first grade *Investigations* curriculum are designed so that students working on different aspects of a mathematical idea can participate fully.

Modifying activities is often as simple as changing the numbers in a problem, or changing the number of turns or another variable in a game. You will find such suggestions, based on the variations that have been tried in classrooms, in each unit. These, along with your own careful listening, observing, and questioning, will provide the information you need to help each student work on significant mathematical ideas at the right level of challenge for that student's development.

# Games: The Importance of Playing More Than Once

Games are used throughout the *Investigations* curriculum as a vehicle for engaging students in important mathematical ideas. The game format is one that most students enjoy, so the potential for repeated experiences with a concept or skill is great. Because most games involve at least two players, students are likely to learn strategies from each other whether they are playing cooperatively or competitively.

The more students play a mathematical game, the more opportunities they have to practice important skills and to think and reason mathematically. The first time or two that students play, they focus on learning the rules. Once they have mastered the rules, their interest turns to the mathematical content. For example, when students play Double Compare, they practice counting, combining, and comparing quantities. Over time, they become familiar with number combinations through frequent experience, rather than by rote memorization.

For many students, repeated experiences with Double Compare lead quite naturally to developing more efficient strategies for combining numbers, to reasoning about numbers and number combinations, and to exploring relationships among number combinations. Later in this unit and in other units in the Investigations curriculum, students will play games involving strategy. Once they are familiar with the rules, they can then begin to attend to strategic questions: *What's the best move I can make? How will my move affect the other player?*

Students need many opportunities to play mathematical games, not just during math time, but at other times as well: in the early morning as students arrive, during indoor recess, or as choices when other work is finished. Games played as homework can be a wonderful way of communicating with parents. Do not feel limited to those times when games are specifically suggested as homework in the curriculum; some teachers send home games even more frequently. One teacher made up "game packs" for loan, placing directions and needed materials in resealable plastic bags, and used these as homework assignments throughout the year. Students often checked out game packs to take home, even on days when homework was not assigned.

# *Using the Calculator in First Grade*

Increasingly sophisticated calculators are used everywhere, from homes to high school classrooms to the workplace. If students are forbidden to use calculators in school while they see adults using them outside of school, they learn that "school" mathematics is nothing like mathematics in the real world. In the world around them, using a calculator is part of real life. We believe that students at all levels need to learn how to use calculators effectively and appropriately as a tool, just as throughout the elementary grades they learn to interpret maps, measure with rulers, and use coins.

Calculators enable students at all levels to apply their reasoning and problem-solving skills to a wider variety of problems. Students can combine the use of calculators with mental calculation or work with manipulatives as they solve problems that use large numbers or require many calculations.

For example, in the middle of the year in one first grade class, students were asked to determine how many crackers the teacher had brought in for snack: there were eight bags of crackers, and each contained six crackers. Of the many methods students came up with, several involved combining the use of the calculator with other mathematical tools.

■ One student recorded 6 + 6 + 6 + 6 + 6 + 6 + 6 + 6, explaining that each 6 represents the number of crackers in one bag, and then used a calculator to find the sum.

■ Another student also began by recording 6 + 6 + 6 + 6 + 6 + 6 + 6 + 6, but was able to solve more of the problem in her head. She knew that 6 + 6 is 12, and recorded 12 under each pair of sixes:

$$6 + 6 + 6 + 6 + 6 + 6 + 6 + 6$$
$$12 + 12 + 12 + 12$$

She then used the calculator to add the four twelves.

■ Another student used the calculator to help her solve the problem by successive "doubling." She explained that since there are six in one bag, there are 12 in two bags. She knew that doubling 12 would give her the number in four bags, and doubling the number in four bags would give her the total in all eight. She then used the calculator to add 12 and 12 to get 24, and then to add 24 and 24 to get 48.

■ Yet another student recognized the problem as a multiplication situation. He knew that the problem could be represented by 6 × 8, but did not know how much that was. He used the calculator to find 6 × 8. Then, to check, he built a representation eight towers of six interlocking cubes, and counted all the cubes.

**When Should Students Use Calculators?** You will probably find many situations in which using calculators can facilitate students' work with large numbers, as it did for the bags of crackers problem. Some of these situations may arise during math time, others outside of math (for example, finding the total number of cans collected for a recycling project, the amount of money collected at a bake sale, the number of cookies needed for a class party).

You will find also many situations in which you do *not* want students to use calculators. When students are exploring number combinations with the How Many of Each? problems, and when they are developing strategies for solving story problems, we suggest that students focus on strategies built on what they know about counting and number relationships.

As students progress through the elementary grades, they will begin making their own choices about when it is appropriate to combine calculators with their own reasoning and when it is appropriate to use other tools, such as mental calculation and estimation, paper and pencil, and manipulatives. Students should have opportunities throughout grade 1 to freely explore and experiment with all these tools.

# Counting

Counting is an important focus in the grade 1 *Investigations* curriculum, as it provides the basis for much of mathematical understanding. As students count, they are learning how our number system is constructed, and they are building the knowledge they need to begin to solve numerical problems. They are also developing critical understandings about how numbers are related to each other and how the counting sequence is related to the quantities they are counting.

Counting routines can be used to support and extend the counting work that students do in the *Investigations* curriculum. As students work with counting routines, they gain regular practice with counting in familiar classroom contexts, as they use counting to describe the quantities in their environment and to solve problems based on situations that arise throughout the school day.

## How Many Are Here Today?

Since you must take attendance every day, this is a good time to look at the number of students in the classroom in a variety of ways.

Ask students to look around and make an estimate of how many are here today. Then ask them to count.

At the beginning of the year, students will probably find the number at school today by counting each student present. To help them think about ways to count accurately, you can ask questions like these:

**How do we know we counted accurately? What are different methods we could use to keep track and make sure we have an exact count?** (For example, you could count around a circle of seated students, with each student in turn saying the next number. Or, all students could start by standing up, then sit down in turn as each says the next number.)

**Is there another way we could count to double-check?** (For example, if you counted around the circle one way, you could count around the circle the other way. If you are using the standing up/sitting down method, you could recount in a different order.)

You might want to count at other times of the day, too, especially when several students are out of the room. For example, suppose groups of students are called to the nurse's office for hearing examinations. Each time a new group of students leaves, you might ask the class to look around and think about how many students are in the room now:

**So, this time Diego's table and Mia's table both went to the nurse. Usually we have 28 students here. Look around. What do you think? Don't count. Just tell me about how many students might be here now. Do you think there are more than 5? more than 10? more than 20?**

Later in the year, some students may be able to use some of the information they know about the total number of students in the class and how many students are absent to reason about the number present. For example, suppose 26 students are in class on Monday, with 2 students absent. On Tuesday, one of those students comes back to school. How many students are in class today? Some students may still not be sure without counting from one, but other students may be able to reason by counting on or counting back, comparing yesterday and today. For example, a student might solve the problem in this way:

> "Yesterday we had 26 students, and Michelle and Chris were both absent. Today, Chris came back, so we have one more person, so there must be 27 today."

Another might solve it this way:

> "Well we have 28 students in our class when everyone's here. Now only Michelle is absent, so it's one less. So it's 27."

From time to time, you might keep a chart of attendance over a week or so, as shown below. This helps students become familiar with different combinations of numbers that make the same total. If you have been doing any graphing, you might want to present the information in graph form.

| Day | Date | Present | Absent | Total |
|---|---|---|---|---|
| Monday | March 2 | 26 | 2 | 28 |
| Tuesday | March 3 | 27 | 1 | 28 |
| Wednesday | March 4 | 27 | 1 | 28 |
| Thursday | March 5 | 27 | 1 | 28 |
| Friday | March 6 | 28 | 0 | 28 |
| Monday | March 9 | 28 | 0 | 28 |
| Tuesday | March 10 | 26 | 2 | 28 |
| Wednesday | March 11 | 25 | 3 | 28 |

After a week or two, look back over the data you have collected. Ask questions about how things have changed over time.

**In two weeks of attendance data, what changes? What stays the same?**

**On which day were the most students here? How can you tell? Which day shows the least students here? What part of the [chart] gives you that information?**

Another idea (for work with smaller numbers) is to keep track of the number of girls and boys present and absent each day. Again, many students will count by 1's. Later in the year, some will also reason about these numbers:

> "There are two people absent today and they're both girls. We usually have 14 girls, and Kaneisha's sick, that's 13, and Claire's sick, that's 12."

## Can Everyone Have a Partner?

Attendance can be an occasion for students to think about making groups of two:

**We have 26 students here today. Do you think that everyone can have a partner if we have 26 students?**

Students can come up with different strategies for solving this problem. They might draw 26 stick figures, then circle them in 2's. They could count out 26 cubes, then put them together in pairs. They might arrange themselves in 2's, or count by 2's.

At the beginning of the year, many of your students will need to count by 1's from the beginning each time you add two more students, but gradually some will begin to notice which numbers can be broken up into pairs:

> "I know 13 doesn't work, because you can do it with 12, and 13's one more, so you can't do it."

Some students will begin to count by 2's, at least for the beginning of the counting sequence. Then, as the numbers get higher, they may still be able to keep track of the 2's, but need to count by 1's:

> "So, that's 2, 4, 6, 8, 10, 12, um, 13, 14 . . . 15, 16."

As you explore 2's with your students, keep in mind that many of them will need to return to 1's as a way to be sure. Even though some students learn the counting sequence 2, 4, 6, 8, 10, 12 . . . by rote, they may not connect this counting sequence to the quantities it represents at each step.

One teacher found a way to help students develop meaning for counting by 2's. She took photographs of each student, backed them with cardboard, then used them during the morning meeting as a model for making pairs. She laid out the photos in two columns, and asked about the new total after the addition of each pair:

We have 10 photos out so far. The next two photos are for William and Yanni. When we put those two photos down, how many photos will we have?

Lining up is another time to explore making pairs. Before lining up, count how many students are in class (especially if it's different from when you took attendance). Ask students whether they think the class will be able to line up in even pairs. For many first grade students, the whole class is too many people to think about. You can ask about smaller groups:

What if Kristi Ann's table lines up first? Do you think we could make even partners with the people at that table?

What about Shavonne's table? … Do you think Shavonne's table will have an extra? How do you know?

Is there another table that would have an extra that we could match up with the extra person from Shavonne's table?

Once students are lined up in pairs, they can count off by 2's. Because most first graders will need to hear all the numbers to keep track of how the counting matches the number of people, ask them to say the first number in the pair softly and the second one loudly. Thus the first pair in line can say, "1, 2," the second pair can say, "3, 4," and so forth.

## Counting to Solve Problems

Be alert to classroom activities that lend themselves to a regular focus on solving problems through counting. Use these situations as contexts for counting and keeping track, estimating small quantities, breaking quantities into parts, and solving problems by counting up or back. For example, take a daily milk count:

Everyone who is buying milk today stand up. Without counting yet, who has an idea how many students might be standing up? Is it more than 5? more than 10? more than 50? … Now, let's count. How could we keep track today so that we get an accurate count?

You can make a problem out of lunch count:

We found out that 23 students are buying school lunch today. We have 27 students here. So how many students brought their own lunch from home today?

Watch for the occasional sharing situation:

Claire brought in some cookies she made to share for snack. She brought 36 cookies. Is that enough for everyone to have one cookie, including me and our student teacher? Oh, and Claire wants to invite her little brother to snack. Do we have enough for him, too? Will there be any cookies left over?

The sharing of curriculum materials can also be the basis of a problem:

Each pair of students needs a deck of number cards to share. While I'm getting things together, work on this problem with your partner. We said this morning that we have 26 students here. If I need one deck for each pair, how many decks do I need?

# Exploring Data

Through data routines at grade 1, students gain experience working with categorical data—information that falls into categories based on a common feature (for example, a color, a shape, or a shared function). The data routines specifically extend work students do in the *Investigations* curriculum. The Guess My Rule game and its many variations (introduced in the unit *Survey Questions and Secret Rules*) can be used throughout the year for practice with organizing sets into categories and finding ways to describe those categories—a fundamental part of analyzing data. Students can also practice collecting and organizing categorical data with quick class surveys that focus on their everyday experiences; this practice supports the survey-taking they do in the curriculum.

## Guess My Rule

Guess My Rule is a classification game in which players try to figure out the common characteristic, or attribute, of a set of objects. To play the game, the rule maker (who may be the teacher, a student, or a small group) decides on a secret rule for classifying a particular group of things. For example, a rule for classifying people might be WEARING STRIPES.

The rule maker (always the teacher when the game is first being introduced) starts the game by giving some examples of objects or people who fit the rule. The guessers then try to find other items that fit the same rule. Each item (or person) guessed is added to one of two groups—either *does fit* or *does not fit* the rule. Both groups must remain clearly visible to the guessers so they can make use of all the evidence as they try to figure out the rule.

Emphasize to the players that "wrong" guesses are as important as "right" guesses because they provide useful clues for finding the rule. When you think most students know the rule, ask for volunteers to share their ideas with the class.

Once your class is comfortable with the activity, students can choose the rules. Initially, you may need to help students choose appropriate rules.

**Guess my Rule with People** When sorting people according to a secret rule, always base the rule on just one feature that is clearly visible, such as WEARING A SHIRT WITH BUTTONS, or WEARING BLUE. When students are choosing the rule, they may choose rules that are too obvious (such as BOY/GIRL), so vague as to apply to nearly everyone (WEARING DIFFERENT COLORS), or too obscure (HAS AN UNTIED SHOELACE). Guide and support students in choosing rules that work.

**Guess My Rule with Objects** Class sets of attribute blocks (blocks with particular variations in size, shape, color, and thickness) are a natural choice for Guess My Rule. You can also use collections of objects, such as sets of keys, household container lids, or buttons. One student sorts four to eight objects according to a secret rule. Others take turns choosing an object from the collection that they think fits the rule and placing it in the appropriate group. If the object does not fit, the rule maker moves it to the NOT group. After several objects have been correctly placed, students can begin guessing the rule.

**Guess My Object** Once students are familiar with Guess My Rule, they can use the categories they have been identifying to play another guessing game that also involves thinking about attributes. In this routine, students guess, by the process of elimination, which particular one of a set of objects has been secretly chosen. This works well with attribute blocks or object collections.

To start, place about 20 objects where everyone can see them. The chooser secretly selects one of the objects on display, but does not tell which one (you may want the chooser to tell you, privately). Other students ask yes-or-no questions, based on attributes, to get clues to help them identify the chosen object. After each answer, students move to one side the objects that have been eliminated. That is, if someone asks "Is it round?" and the answer is yes, all objects that are *not* round are moved aside.

Pause periodically to discuss which questions help eliminate the most objects. For example, "Is it this one?" eliminates only one object, whereas "Is it red?" may eliminate several objects. For more challenge, students can play with the goal of identifying the secret object with the fewest questions.

## Quick Surveys

Class surveys can be particularly engaging when they connect to activities that arise as a regular part of the school day, and they can be used to help with class decisions. As students take surveys and analyze the results, they get good practice with collecting, representing, and interpreting categorical data.

Early in first grade, to keep the surveys quick and the routine short, use questions that have exactly two possible responses. For example:

**Would you rather go outside or stay inside for recess today?**

**Will you drink milk with your lunch today?**

**Do you need left-handed or right-handed scissors?**

As the school year progresses, you might include some survey questions that are likely to have more than two responses:

**Which of these three books do you want me to read for story time?**

**Who was your teacher last year?**

**Which is your favorite vegetable growing in our class garden?**

**How old are you?**

**In which season were you born?**

Try to choose questions with a predictable list of just a few responses. A question like "What is your favorite ice cream flavor?" may bring up such a wide range of responses that the resulting data is hard to organize and analyze.

As students become more familiar with classroom surveys, invite the class to brainstorm questions with you. You may decide to avoid survey questions about sensitive issues such as families, the body, or abilities, or you might decide to use surveys as a way of carefully raising some of these issues. In either case, it is best to avoid questions about material possessions ("Does your family have a car?").

Once the question is chosen, decide how to collect and represent data. Be sure to vary the approach. One time, you might collect data by recording students' responses on a class list. Another time, you might take a red interlocking cube for each student who makes one response, a blue cube for each student who makes the other response. Another time, you might draw pictures. If you have prepared Kid Pins and survey boards for use in *Mathematical Thinking at Grade 1,* these can be used for collecting the data from quick surveys all year.

Initially, you may need to help students organize the collected data, perhaps by stacking cubes into "bars" for a "graph," or by making a tally. Over time, students can take on more responsibility for collecting and organizing the data.

Always spend a little time asking students to describe, compare, and interpret the data.

**What do you notice about these data?**

**Which group has the most? the least? How many more students want [recess indoors today]?**

**Why do you suppose more would rather [stay inside]? Do you think we'd get similar data if we collected on a different day? What if we did the same survey in another class?**

# Understanding Time and Changes

These routines help students develop an understanding of time-related ideas such as sequencing of events, understanding relationships among time periods, and identifying important times in their day.

Young students' understanding of time is often limited to their own direct experiences with how important events in time are related to each other. For example, explaining that an event will occur *after* a child's birthday or *before* an important holiday will help place that event in time for a child. Similarly, on a daily basis, it helps to relate an event to a benchmark time, such as *before* or *after* lunch. Both calendars and daily schedules are useful tools in sequencing events over time and preparing students for upcoming events. These routines help young students gain a sense of basic units of time and the passage of time.

## Calendar

The calendar is a real-world tool that people use to keep track of time. As students work with the calendar, they become more familiar with the sequence of days, weeks, and months, and the relationships among these periods of time. Calendar activities can also help students become more familiar with relationships among the numbers 1–31.

**Exploring the Monthly Calendar**  At the start of each month, post the monthly calendar and ask students what they notice about it. Some students might focus on arrangement of numbers or total number of days, while others might note special events marked on the calendar, or pictures or designs on the calendar. All these kinds of observations help students become familiar with time and ways that we keep track of time. You might record students' observations and post them near the calendar.

As the year progresses, encourage students to make comparisons between the months. Post the calendar for the new month next to the calendar for the month just ending and ask students to share their ideas about how the two calendars are similar and different.

**Months and Years**  To help students see that months are part of a larger whole, display the entire calendar year on a large sheet of paper. Cut a small calendar into individual monthly pages and post the sequence of months on the wall. You might decide to post the months according to the school year, September through August, or the calendar year January through December. At the start of each month, ask students to find the position of the new month on the larger display. From time to time, you might also use this display to point out dates and distances between them as you discuss future events or as you discuss time periods that span a month or more. (Last week was February vacation. How many weeks until the next vacation?)

**How Many More Days?**  Ask students to figure out how long until special events, such as birthdays, vacations, class trips, holidays, or future dates later in the month. For example:

**Today is October 5. How many more days until October 15?**

**How many more days until [Nathan's] birthday?**

**How many more days until the end of the month?**

Ask students to share their strategies for finding the number of days. Initially, many students will count each subsequent day. Later, some students may begin to find their answers by using their growing knowledge of calendar structure and number relationships:

> "I knew there were three more days in this row and I added them to the three days in the next row. That's 6 more days."

Others may begin using familiar numbers such as 5 or 10 in their counting:

> "Today is the 5th. Five more days is 10, and five more is 15. That's 10 more days until October 15."

For more challenge, ask for predictions that span two calendar months. For example, you might post the calendar for next month along side of the calendar of this month and ask a question like this:

**It's April 29 today. How many more days until our class trip on May 6?**

Note that we can refer to a date either as October 15 or as the 15th day of October. Vary the way you refer to dates so that students become comfortable with both forms. Saying "the 15th day of October" reinforces the idea that the calendar is a way to keep track of days in a month.

**How Many Days Have Passed?** Ask questions that focus on events that have already occurred:

**How many days have passed since [a special event]? since the weekend? since vacation?**

**Mixed-Up Dates** If your monthly display calendar has date cards that can be removed or rearranged, choose two or three dates and change their position on the calendar so that the numbers are out of order. Ask students to fix the calendar by pointing out which dates are out of order.

Groups of two or three can play this game with each other during free time. Students can also remove all the date cards, mix them up, and reassemble the calendar in the correct order. You might mark the space for the first day of the month so that students know where to begin.

## Daily Schedule

The daily schedule narrows the focus of time to hours and shows students the order of familiar events over time. Working with schedules can be challenging for many first graders, but regular opportunities to think and talk about the idea will help them begin predicting what comes next in the schedule. They will also start to see relationships between particular events in the schedule and the day as a whole.

**The School Day** Post a schedule for each school day. Identify important events (start of school, math, music, recess, reading, lunch) using pictures or symbols and times. Include both analog (clock face) and digital (10:15) representations. Discuss the daily schedule each day with students using words such as *before* math, *after* recess, *during* the morning, *at the end of* the school day. Later in the school year you can begin to identify the times that events occur as a way of bridging the general idea of sequential events and the actual time of day.

**The Weekend Day** Students can create a daily schedule, similar to the class schedule, for their weekend days. Initially they might make a "timeline" of their day, putting events in sequential order. Later in the year they might make another schedule where they indicate the approximate time of day that events occur.

## Weather

Keeping track of the weather engages young students in a real-life data collection experience in which the data they collect changes over time. By displaying this ongoing collection of data in one growing representation, students can compare changes in weather across days, weeks, and months, and observe trends in weather patterns, many of which correspond to the seasons of the year.

**Monthly Weather Data** With the students, choose a number of weather categories (which will depend on your climate); they might include sunny, cloudy, partly cloudy, rainy, windy, and snowy.

If you vary the type of representation you use to collect monthly data, students get a chance to see how similar information can be communicated in different ways. On the following page you'll see some ways of representing data that first grade teachers have used.

At the end of each month (and periodically throughout the month), ask questions to help students analyze the data they are collecting.

Another approach over the entire year is to prepare 10-by-10 grids from 1-inch graph paper, making one grid for each weather category your class has chosen. Post the grids, labeled with the identifying weather word. Each day, a student records the weather by marking off one square on one or more grids; that is, on a sunny day, the student marks a square on the "sunny" grid, and if it's also windy, he or she marks the "windy" grid, too.

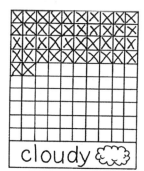

From time to time, students can calculate the total number of days in a certain category by counting the squares. Because these are arranged in a 10-by-10 grid, some students may use the rows of 10 to help them calculate the total number of days. ("That's 10, and another 10 is 20, and 21, 22, 23.")

**Making Weather Representations** After students have had some experience collecting and recording data in the grade 1 curriculum (especially in *Survey Questions and Secret Rules*), they can make their own representation of the weather data. For one month, record the weather data on a piece of chart paper (or directly on your monthly calendar), without organizing it by category. At the end of the month, ask students to total the number of sunny days, rainy days, and so forth, and post this information (perhaps as a tally). Students then make their own representation of the data, using pictures, numbers, words, or a combination of these. Encourage them to use clear categories and show the number of days in each.

Weather data can be collected on displays like these. In the second example, a student draws each day's weather on an index card to add to the graph. The third example uses stick-on dots.

**What is this graph about?**

**What does this graph tell us about the weather this month (so far)?**

**What type of weather did we have for the most days? What type of weather did we hardly ever have?**

**How is the weather this month different from the weather last month? What are you looking at on the graph to help you figure that out?**

**How do you think the weather graph for next month will look?**

**Yearly Weather Data** If you collect and analyze weather data for some period of time, consider extending this over the entire school year. Save your monthly weather graphs, and periodically look back to see and discuss the changes over longer periods of time.

The following activities will help ensure that this unit is comprehensible to students who are acquiring English as a second language. The suggested approach is based on *The Natural Approach: Language Acquisition in the Classroom* by Stephen D. Krashen and Tracy D. Terrell (Alemany Press, 1983). The intent is for second-language learners to acquire new vocabulary in an active, meaningful context.

Note that *acquiring* a word is different from *learning* a word. Depending on their level of proficiency, students may be able to comprehend a word upon hearing it during an investigation, without being able to say it. Other students may be able to use the word orally, but not read or write it. The goal is to help students naturally acquire targeted vocabulary at their present level of proficiency.

We suggest using these activities just before the related investigations. The activities can also be led by English-proficient students.

### compare, more, less

1. Give two cubes, four cubes, and ten cubes to three students. Explain:

   **We are going to compare how many cubes you have.**

   **Tuan** *[with ten]* **has more than Luis and Yukiko.**

   **Luis** *[with two]* **has less than Yukiko and Tuan.**

2. Tell the three students to exchange any number of their cubes with each other.

   **Now let's compare again. Who has more?**

   **Who has less?**

3. Draw three dots on the board. Ask for a volunteer to draw *more* dots next to yours. Ask another volunteer to draw *less*.

### container, holds, full

1. Show an empty paper cup and a transparent jar. Explain that we can call both of these *containers*. Fill the jar to the top with water. Point to the rim of the jar as you explain that it is a *full* container.

2. Ask:

   **Will the paper cup hold more or less water than the jar?**

3. Pour a little water from the jar into the paper cup, and ask:

   **Is the paper cup full?**

   **Do we need more or less water to make it full?**

   Keep pouring small amounts of water and asking:

   **Is the cup now full?**

4. When the cup is finally full, ask students to compare the amount of water with that remaining in the jar.

   **Look** *[point]* **at how much water is in the jar.**
   **Look** *[point]* **at how much water is in the cup.**
   **Let's compare the jar of water** *[point]* **to the cup of water** *[point]*.
   **Which container holds more water?**
   **Which container holds less water?**
   **Which container is full?**

# Blackline Masters

_____, 19____

# Dear Family,

Our class is beginning a unit called *Building Number Sense*. The emphasis is on understanding how numbers can be made from other numbers: 12 can be made from 6 and 6, from 10 and 2, or from 6, 4, and 2. Being able to take numbers apart and put them back together in different ways forms the basis for a lot of the mathematical work that your child will do in the elementary grades.

The children will play games and solve problems in which they break numbers into parts in different ways. For example, looking at a single pattern of 10 dots, some children see it as two groups of 5, some as five groups of 2, and some as two groups of 4 with 2 extra. As they share their ways of thinking, they learn different ways of breaking 10 into parts.

The class will also be solving addition and subtraction story problems. Our emphasis is on understanding the problem situation, finding your own way to solve the problem, and recording your work to show clearly how you solved it.

Throughout the unit, the children will solve problems in ways that make sense to them. Sometimes they will work out solutions by counting objects. Other times, they may draw pictures, count in their heads or on their fingers, or begin using their growing knowledge of number combinations.

While our class is working on this unit you can help in several ways:

■ For homework, your child will be bringing home math games we play at school. Playing the games frequently will help your child learn. Help your child find a safe place, such as a file folder or manila envelope, to store the game materials, since some of them will be used repeatedly throughout the unit.

■ Have a large collection of small objects, such as buttons, paper clips, or pennies. Your child can use these for working out solutions to problems. You will also need them for some of the games.

■ As your child works on story problems at home, encourage written work that shows his or her thinking. Children can use pictures, numbers, words, or a combination; all are important ways of showing mathematical thinking.

Thank you for your help.

Sincerely,

# Compare Dots

**Materials:** Dot Cards, Set A

**Players:** 2

**Object:** Decide which of two cards has more dots.

## How to Play

1. Mix the cards and deal them evenly to each player. Place your stack of cards facedown in front of you.

2. At the same time, both of you turn over the top card in your stack. Look at the dots. If your card has more dots, you say "Me!" If the two cards are the same, turn over the next card.

3. Keep turning over cards. Each time, say "Me!" if your card has more dots.

4. The game is over when you have both turned over all the cards in your stack.

## Variations

a. If you have **fewer** dots, you say "Me."

b. Play with three people. Look at all three numbers. If you have the most dots, say "Me."

c. Play the game Double Compare Dots. Turn over the top two cards in your pile each time. Find the total number of dots on your two cards. The player with the higher total says "Me."

# Copying Counters

**Note to Families**
For counters, you can use buttons, pennies, paper clips, beans, or toothpicks.

**Materials:** About 60 counters

**Players:** 2

**Object:** Make an exact copy of your partner's pattern or design.

## How to Play

1. Both players make their own pattern or design from about 15 counters.

2. When both of you have finished, make an exact copy of what your partner has made. (If you are using colored counters, your colors do not need to match.)

3. When the copies are done, check that your partner's copy uses the same number of counters as yours. Also check that it is the exact same size and shape as yours.

## Variation

Use a different number of counters in your design, such as 10 or 25.

# Counting Pattern Blocks

Block            Number

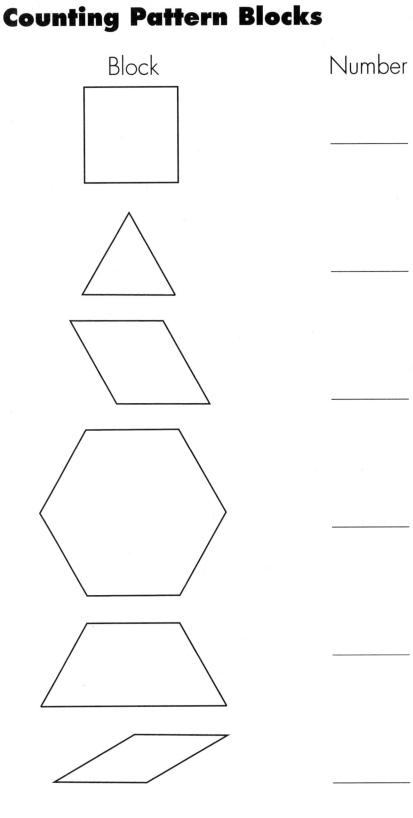

_____

_____

_____

_____

_____

_____

Total number of blocks    _____

# Making Quick Images

Number of dots: _____

# DOT CARDS, SET A

You need four copies of this sheet to make a complete set of 32 cards.

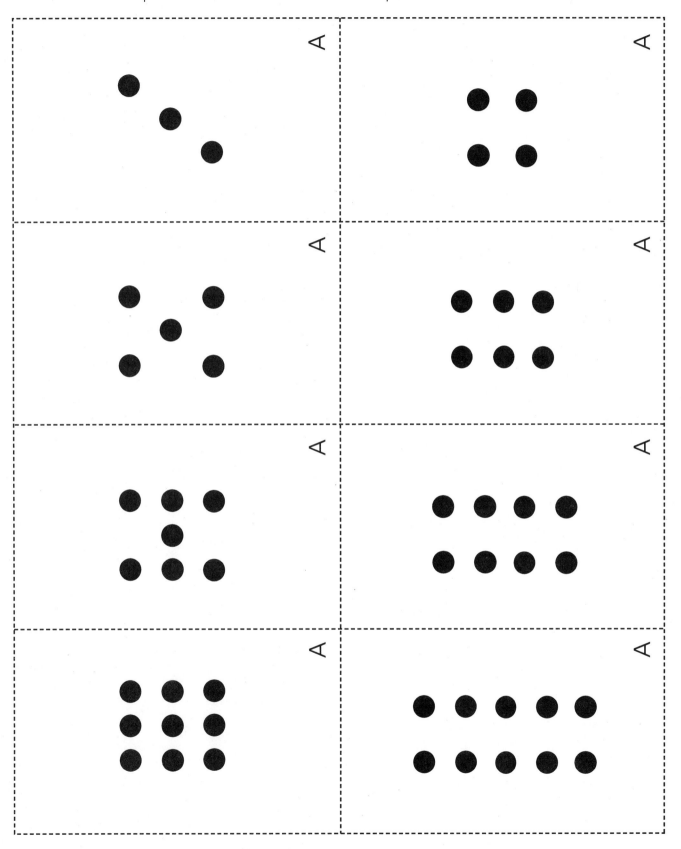

# DOT CARDS, SET B

You need four copies of this sheet to make a complete set of 32 cards.

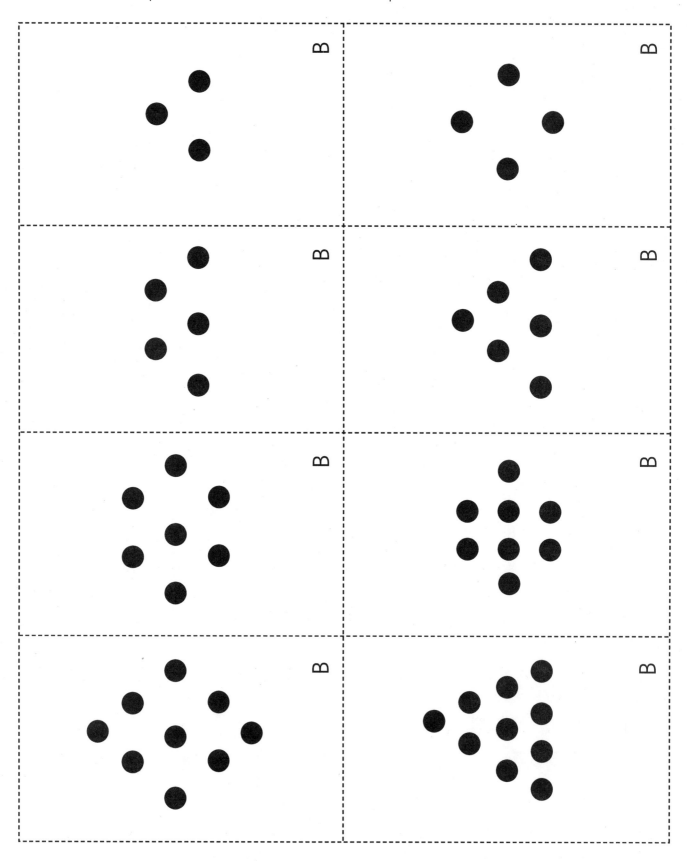

*Investigation 1 Resource*
*Building Number Sense*

# DOT CARDS, SET C

You need four copies of this sheet to make a complete set of 32 cards.

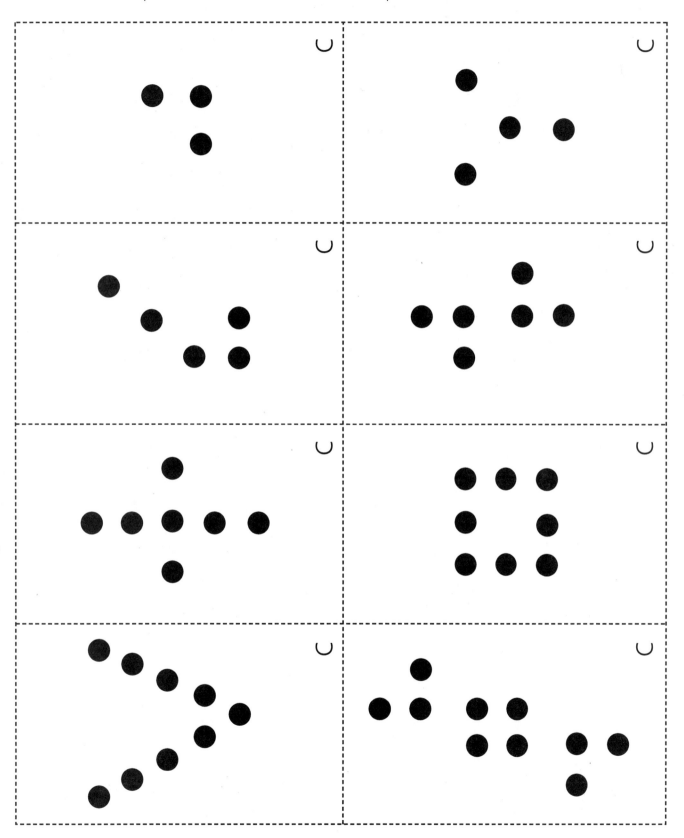

# DOT CARDS, SET D

You need four copies of this sheet to make a complete set of 40 cards.

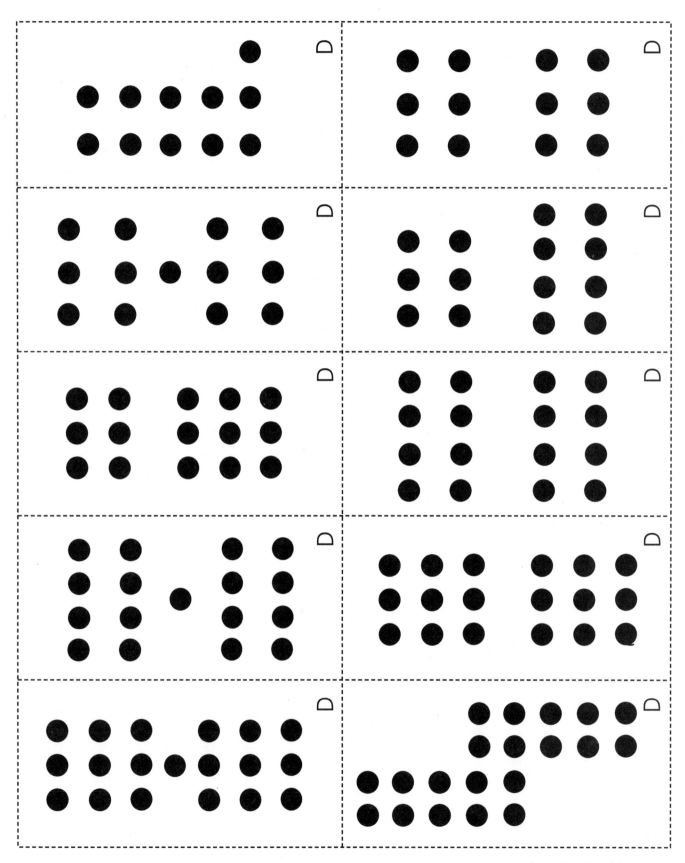

# EASY AND HARD QUICK IMAGES

Make a transparency of these three images. Cut apart.

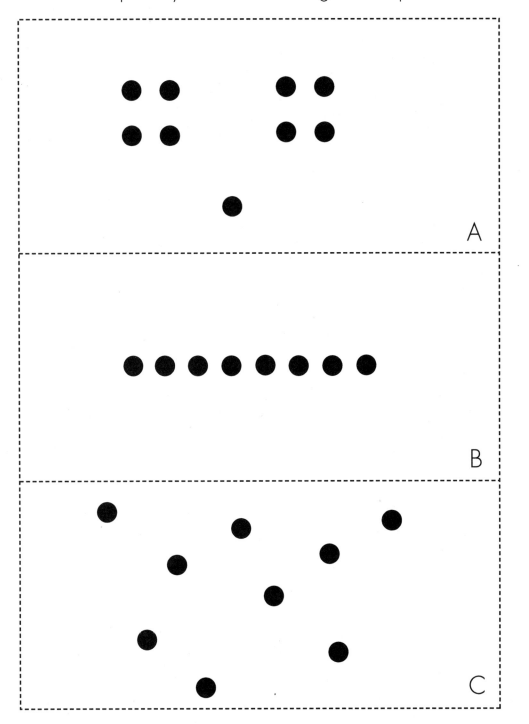

A

B

C

# Turtles and Frogs

I have ___11___ turtles and frogs.

How many of each could I have?

Keep track of your work. You can use pictures, numbers, or words.

# Double Compare

**Materials:** Deck of Number Cards 0–10
(remove the wild cards)

**Players:** 2

**Object:** Decide which of two totals is greater.

**Note to Families**
In this game, your child will be finding the totals of pairs of numbers. You will need a set of Number Cards to play this game.

## How to Play

1. Mix the cards and deal them evenly to each player. Place your stack of cards facedown in front of you.

2. At the same time, both of you turn over the top two cards in your stack. Look at your two numbers and find the total. Then find the total of the other player's numbers.

   If your total is more than the other player's, say "Me!" If the two totals are the same, turn over the next two cards.

3. Keep turning over two cards. Say "Me!" each time your total is more.

4. The game is over when you have both turned over all the cards in your stack.

## Variations

a. If your total is **less,** say "Me."

b. Play with three people. Find all three totals. If yours is the most, say "Me."

c. Add the four wild cards to the deck. A wild card can be any number.

# On and Off

**Materials:** Counters (8–12)
On and Off game grid
Sheet of paper

**Players:** 1–3

**Object:** Toss counters over a sheet of paper. Record how many land on and off the paper.

## How to Play

1. Decide how many counters you will toss each time. Write this total number on the game grid.

2. Lay the sheet of paper on a flat surface.

3. Hold the counters in one hand and toss them over the paper.

4. On the game grid, write how many landed on the paper and off the paper.

5. Repeat steps 3 and 4 until you have filled the game grid (eight tosses).

## Optional

Your filled game grid shows different ways to break the total number into two parts. Can you find a way that is not shown?

# On and Off Game Grid

## Game 1

Total number: _____

| On | Off |
|----|-----|
|    |     |
|    |     |
|    |     |
|    |     |
|    |     |
|    |     |
|    |     |
|    |     |

## Game 2

Total number: _____

| On | Off |
|----|-----|
|    |     |
|    |     |
|    |     |
|    |     |
|    |     |
|    |     |
|    |     |
|    |     |

# Counters in a Cup

**Note to Families**
For counters, you can use buttons, pennies, paper clips, beans, or toothpicks. Hide them under any container that you cannot see through. If you do not have a copy of the game grid, write the numbers in two columns on any paper.

**Materials:** Counters (5–10)

Counters in a Cup game grid

Paper cup

**Players:** 2

**Object:** Figure out how many of a set of counters are hidden.

## How to Play

1. Decide how many counters to use each time. Write this total number on the game grid.

2. Player A hides a secret number of counters under the cup and leaves the rest out.

3. Player B figures out how many are hidden and says the number. Lift the cup to check.

4. On the game grid, write the number hidden in the cup and the number left out.

5. Players switch roles. Hide a different number of counters. (It's OK to hide the same number of counters more than once in a game.)

6. Repeat steps 2–5 until you have filled the game grid. (Hide the counters eight times.)

## Optional

Your filled game grid shows different ways to break the total number into two parts. Can you find a way that is not shown?

# Counters in a Cup Game Grid

Total number: _____

| In | Out |
|----|-----|
|    |     |
|    |     |
|    |     |
|    |     |
|    |     |
|    |     |
|    |     |
|    |     |

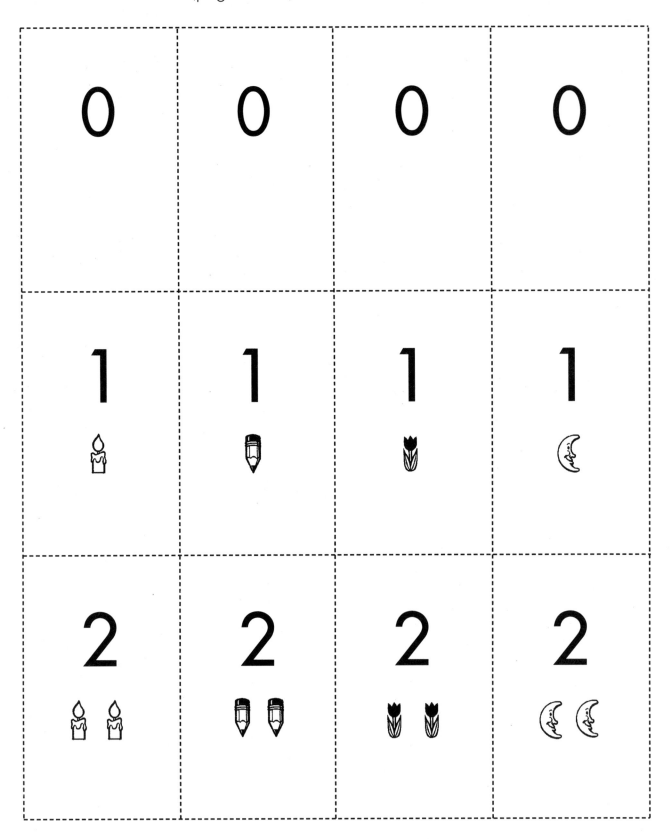

3

3

3

3

4

4

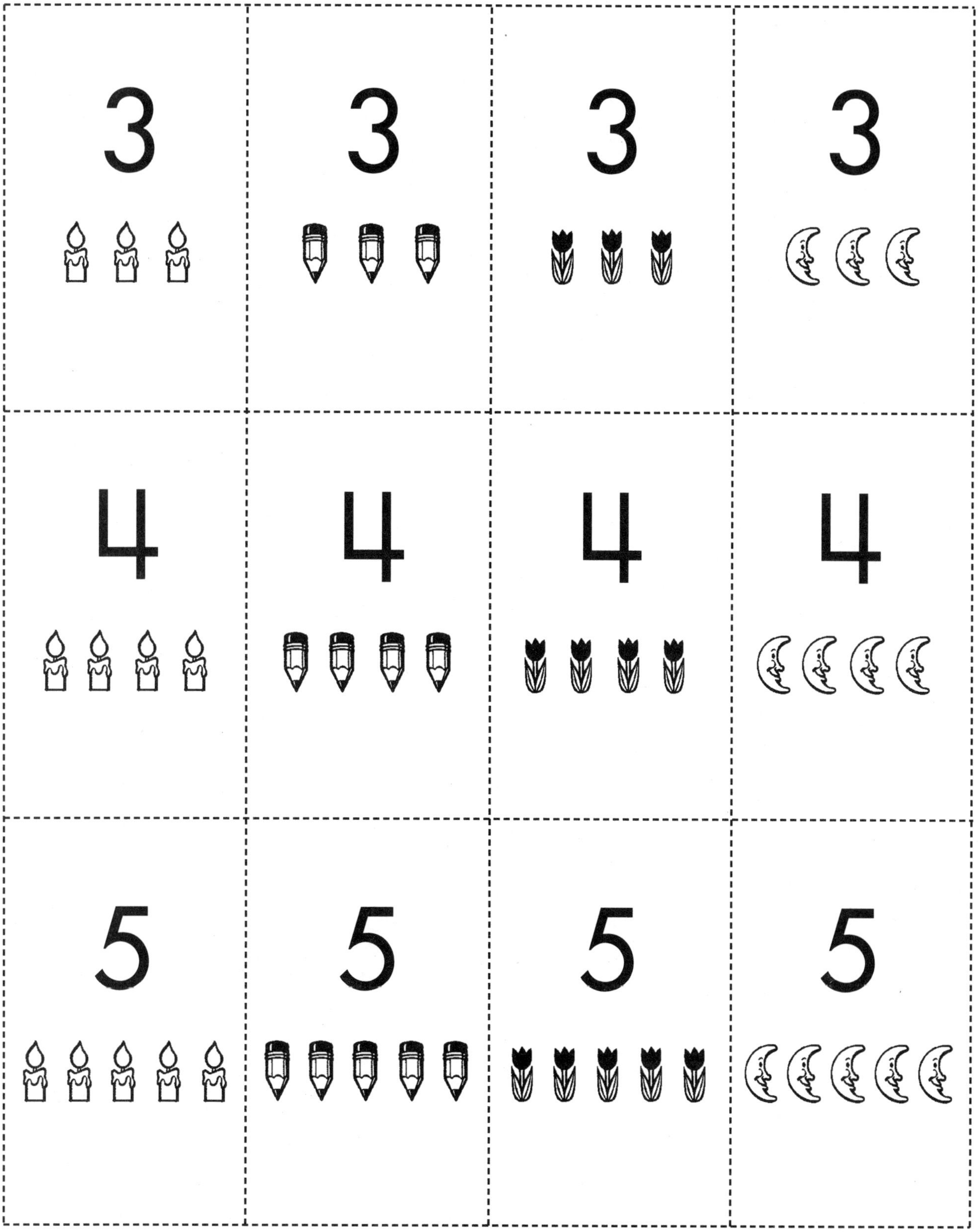

4

4

5

5

5

5

| 6 | 6 | 6 | 6 |
|---|---|---|---|
| 7 | 7 | 7 | 7 |
| 8 | 8 | 8 | 8 |

| 9 | 9 | 9 | 9 |
|---|---|---|---|

| 10 | 10 | 10 | 10 |
|---|---|---|---|

| Wild Card | Wild Card | Wild Card | Wild Card |
|---|---|---|---|

# Numbers at Home

Look around your home for numbers on things.
Find a number between 10 and 100.
Write the number. Draw a picture that shows
where you found it, or explain it in words.

# **Which Holds More?**

I used container _____ and container _____ .

Which container holds more? How do you know?




I used container _____ and container _____ .

Which container holds more? How do you know?

# Ten Turns

**Materials:** One number cube
Counters (50–60)
Ten Turns Game Sheet

**Players:** 2

**Object:** With a partner, collect as many counters as you can.

## How to Play

1. Roll the number cube. What number did you roll? Take that many counters to start your collection. Write the number you rolled and the total number you have. (For the first turn, these numbers are the same.)

2. On each turn, roll the number cube and take that many counters. Find the total number of counters you and your partner have together.

3. After each turn, write the number you rolled and the new total.

4. Play for 10 turns.

## Variations

a. Play for fewer turns or more turns.

b. Roll two number cubes on each turn.

c. Instead of a number cube, use the Number Cards for 1 to 6. Mix them and turn up one at a time.

# Ten Turns Game Sheet

Turn 1.   I rolled _____   Now we have _____

Turn 2.   I rolled _____   Now we have _____

Turn 3.   I rolled _____   Now we have _____

Turn 4.   I rolled _____   Now we have _____

Turn 5.   I rolled _____   Now we have _____

Turn 6.   I rolled _____   Now we have _____

Turn 7.   I rolled _____   Now we have _____

Turn 8.   I rolled _____   Now we have _____

Turn 9.   I rolled _____   Now we have _____

Turn 10.  I rolled _____   Now we have _____

# 100 CHART

| 1 | 2 | 3 | 4 | 5 | 6 | 7 | 8 | 9 | 10 |
|---|---|---|---|---|---|---|---|---|----|
| 11 | 12 | 13 | 14 | 15 | 16 | 17 | 18 | 19 | 20 |
| 21 | 22 | 23 | 24 | 25 | 26 | 27 | 28 | 29 | 30 |
| 31 | 32 | 33 | 34 | 35 | 36 | 37 | 38 | 39 | 40 |
| 41 | 42 | 43 | 44 | 45 | 46 | 47 | 48 | 49 | 50 |
| 51 | 52 | 53 | 54 | 55 | 56 | 57 | 58 | 59 | 60 |
| 61 | 62 | 63 | 64 | 65 | 66 | 67 | 68 | 69 | 70 |
| 71 | 72 | 73 | 74 | 75 | 76 | 77 | 78 | 79 | 80 |
| 81 | 82 | 83 | 84 | 85 | 86 | 87 | 88 | 89 | 90 |
| 91 | 92 | 93 | 94 | 95 | 96 | 97 | 98 | 99 | 100 |

# BLANK 100 CHART

|  |  |  |  |  |  |  |  |  |  |
|--|--|--|--|--|--|--|--|--|--|
|  |  |  |  |  |  |  |  |  |  |
|  |  |  |  |  |  |  |  |  |  |
|  |  |  |  |  |  |  |  |  |  |
|  |  |  |  |  |  |  |  |  |  |
|  |  |  |  |  |  |  |  |  |  |
|  |  |  |  |  |  |  |  |  |  |
|  |  |  |  |  |  |  |  |  |  |
|  |  |  |  |  |  |  |  |  |  |
|  |  |  |  |  |  |  |  |  |  |
|  |  |  |  |  |  |  |  |  |  |

*Investigation 3 • Resource*
*Building Number Sense*

# CUBE PATTERN STRIPS

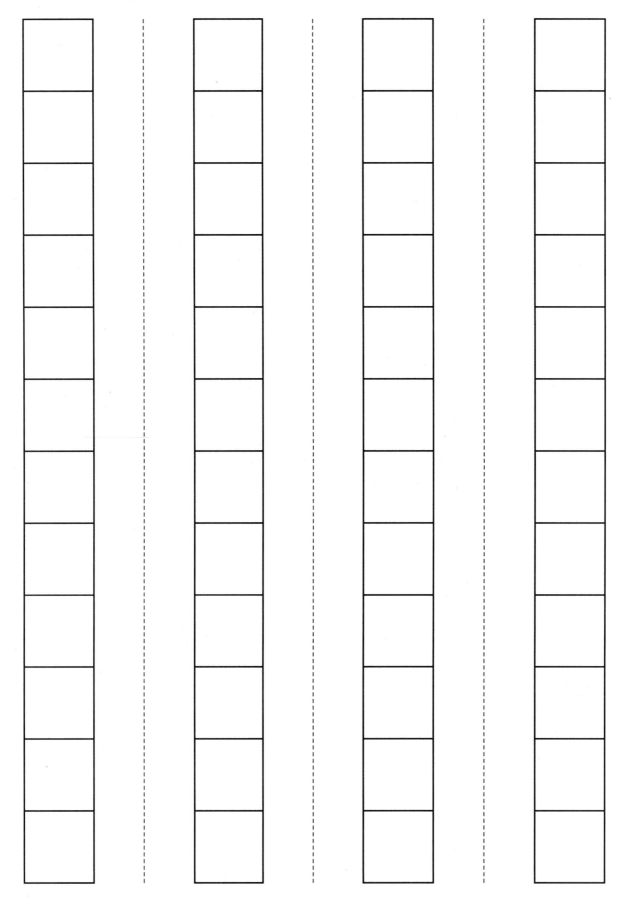

# Apples and Oranges

I went to the store to buy some fruit.
I bought 5 apples and 4 oranges.
How many pieces of fruit did I buy?

Don't write only the answer.
Show how you solved the problem.
Use words, pictures, or numbers.

# Eating Apples

We had 11 apples in a bowl.
We ate 3 of them.
How many apples do we have now?

Don't write only the answer.
Show how you solved the problem.
Use words, pictures, or numbers.

# Five-in-a-Row

**Materials:** Gameboard, counters (about 20)
two number cubes (or, number
cards 1– 6)

**Players:** 2

**Object:** Cover five squares in a row horizontally,
vertically, or diagonally.

## How to Play

1. Roll the number cubes (or turn up two Number
   Cards).

2. Find the sum of the rolls. Look on the board for a
   square that matches the sum. Place a counter on
   that square.

3. Continue in this way until you have covered
   five squares in a row. If no move can be made
   (the sum is covered), roll again.

## Variations

a. You and your partner have your own boards.
   The game is over when both of you have five
   in a row.

b. Roll three number cubes on each turn. Make
   sums with any two of the numbers rolled.
   For example: You roll 3, 1, and 5. You could
   cover 4 (3 and 1), 6 (5 and 1), or 8 (5 and 3).
   If you and your partner have your own boards,
   you could make different sums to cover.

# Five-in-a-Row Board A

| 2 | 3 | 4 | 5 | 6 |
|---|---|---|---|---|
| 6 | 7 | 7 | 8 | 9 |
| 10 | 11 | 12 | 11 | 10 |
| 9 | 8 | 7 | 7 | 6 |
| 6 | 5 | 4 | 3 | 2 |

# Five-in-a-Row Board B

| | | | | |
|---|---|---|---|---|
| 12 | 11 | 10 | 9 | 8 |
| 7 | 7 | 6 | 5 | 4 |
| 3 | 2 | 2 | 3 | 4 |
| 5 | 6 | 6 | 7 | 7 |
| 8 | 9 | 10 | 11 | 12 |

# Five-in-a-Row Board C

| 2 | 2 | 3 | 3 | 4 |
|---|---|---|---|---|
| 4 | 5 | 5 | 6 | 6 |
| 6 | 7 | 7 | 7 | 7 |
| 8 | 8 | 9 | 9 | 10 |
| 10 | 11 | 11 | 12 | 12 |

# Dot Addition

**Materials:** Dot Addition Cards (20 dot cards), Dot Addition Board, sheet of paper

**Players:** 1–3

**Object:** Use combinations of dot cards to make the numbers on the board.

**Note to Families**
You may use any of several Dot Addition Boards that your child may bring home. Although only four cards fit across the board, it's OK to use more to make a number. Just overlap them or let them go off the board.

## How to Play

1. Lay out the 20 cards faceup in four rows of five. Place the board beside them.

2. Work together to make each number on the Dot Addition Board with your cards.

3. You can't use a card twice. For example, if you use three 3's to make 9, you can't use all 3's to make 12.

4. You may rearrange your cards at any time until you have made all the numbers on your board.

5. At the end of the game, write the number combinations you made on a sheet of paper.

## Variations

a. Play again with the same board. Find a different way to make each number with the cards.

b. Make your own Dot Addition Boards.

# Dot Addition Board A

| 6 |
|---|

| 8 |
|---|

| 10 |
|---|

| 12 |
|---|

# Dot Addition Board B

| 6 |
|---|
|   |

| 9 |
|---|
|   |

| 10 |
|---|
|    |

| 15 |
|---|
|    |

# Dot Addition Board C

**8**

**9**

**12**

**16**

# BLANK DOT ADDITION BOARD

*Investigation 4 Resource*
*Building Number Sense*

# DOT ADDITION CARDS

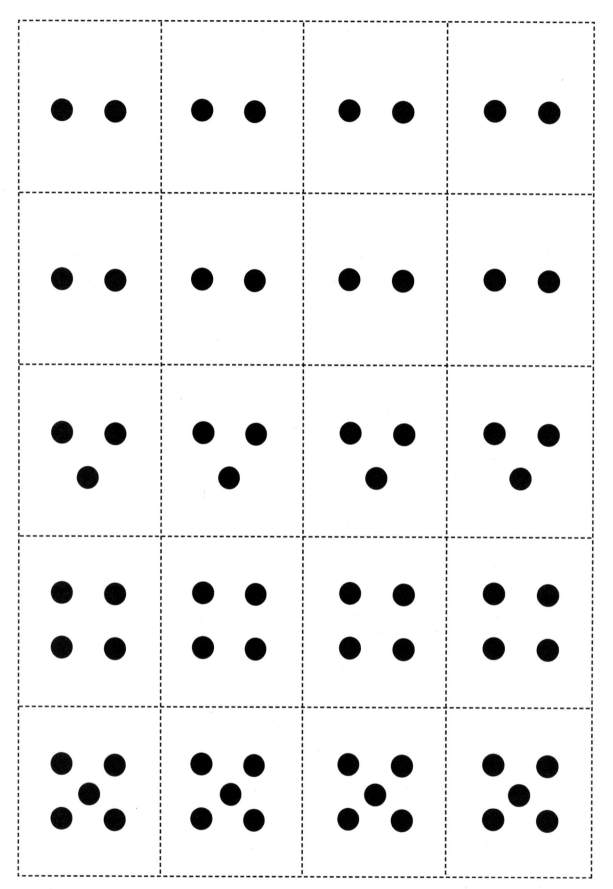

# STORY PROBLEMS, SET A

Copy one set per student. Cut apart and place copies of each problem in a separate envelope.

**1.** Rosa had 6 toy cars. Her mom gave her 6 more. How many toy cars does Rosa have now?

A

**2.** Steve has 5 marbles in his pocket. He has 9 marbles in a bag. How many marbles does he have?

A

**3.** A squirrel ate 8 nuts. Then she ate 7 more nuts. How many nuts did she eat?

A

**4.** Rosa had 11 library books. She took 4 of them back. Now how many books does she have??

A

**5.** Steve and Rosa had 12 apples. They ate 6 of them. How many apples are left?

A

**6.** Rosa had 15 pennies. She spent 6 pennies. How many pennies did she have then?

A

© Dale Seymour Publications®

*Investigation 4 Resource*
*Building Number Sense*

# STORY PROBLEMS, SET B

Copy one set per student. Cut apart and place copies of each problem in a separate envelope.

1. 10 children were playing at the park. Then 8 more came to play. How many children were at the park?

B

2. Kim and Tito picked 11 red flowers. They picked 5 white flowers. How many flowers did they pick?

B

3. Kim has 5 cookies. Tito has 4 cookies. Jill has 3 cookies. How many do they have in all?

B

4. Kim had 16 pennies in her pocket. The pocket had a hole. 8 pennies fell out. Now how many pennies are in her pocket?

B

5. Tito had 14 fish in a tank. He gave away 9 fish. How many fish does he have now?

B

6. Kim saw 20 ducks on the pond. Then 9 ducks flew away. How many ducks were still on the pond?

B

*Investigation 4 Resource*
*Building Number Sense*

# STORY PROBLEMS, SET C (CHALLENGES)

These are optional. Have available for about half the class. Cut apart and place copies of each problem in a separate envelope.

1. Steve and Rosa picked cherries. They ate 18 cherries for lunch. Rosa ate 6 cherries after lunch. How many cherries did they eat?

C

2. Rosa had 11 stamps. Her father gave her 12 more. Now how many stamps does Rosa have?

C

3. 4 children were on the bus. At the next stop, 13 more got on. Then 6 more children got on. How many children were on the bus?

C

4. Steve had 19 pennies. He spent 15 pennies. Now how many does he have?

C

5. Rosa had 21 balloons. She gave 3 of them away. Now how many balloons does she have?

C

6. Rosa's mom had 23 hens. She sold 8 of them. Then how many hens did she have?

C

# STORY PROBLEMS, SET D

Copy one set per student. Cut apart and distribute the problems one at a time.

**1.** Kim made 7 paper hats. Tito made 6 paper hats.
How many paper hats did they make?

D

**2.** Tito had 15 cookies. He gave 9 to his friends.
How many cookies did Tito have then?

D

**3.** Tito found 5 frogs. Kim found 8 frogs. Jill found 7 frogs.
How many frogs did they find?

D

*Investigation 4 Resource*
*Building Number Sense*

# Game Record Sheet

Game:

_____

Play this game at home. You should have the directions and other things you need. After you play, fill out and return this sheet.

Who played the game?

_____

Write about what happened when you played the game.

_____

_____

_____

_____

_____

_____

**Note to Families**

Please play this game with your child, then help fill out and return this sheet. The more times children play a mathematical game, the more practice they get with important skills and with reasoning mathematically.

# Practice Pages

This section provides optional homework for teachers who want or need to give more homework than is suggested to accompany the activities in this unit. With the games or problems included here, students get additional practice in learning about number relationships and solving number problems. Whether or not the *Investigations* unit you are presenting in class focuses on number skills, continued work at home on developing number sense will benefit students. In this unit, practice pages include the following:

**How Many of Each? Problems**  This type of problem is introduced in *Mathematical Thinking at Grade 1,* and students encounter it again in Investigation 2 of this unit. Six additional problems are provided here. You can modify any of the numbers and make up new problems in this format, using numbers that are appropriate for your students. Students need not always work with a different total each time they do a How Many of Each? problem. Repeating the same total with different objects gives them more practice with number combinations.

**Missing Numbers**  Students fill in the missing numbers on 100 charts with some of the numbers left blank. A similar Missing Numbers activity and 100 charts are introduced in Investigation 3 of this unit. Three different 100 charts with missing numbers are provided here. You can also use a blank 100 chart (p. 207) to make up your own Missing Numbers sheets. For extra challenge, you could ask students to fill in an entire blank chart. Briefly introduce this activity in class before sending it home.

## Practice Page A

I have ___10___ circles and squares.

How many of each could I have?

Keep track of your work. You can use pictures, numbers, or words.

# Practice Page B

I have ___12___ dragons and mice.

How many of each could I have?

Keep track of your work. You can use pictures, numbers, or words.

## Practice Page C

I have ___13___ apples and pears.

How many of each could I have?

Keep track of your work. You can use
pictures, numbers, or words.

## Practice Page D

I have ___15___ stars and moons.

How many of each could I have?

Keep track of your work. You can use pictures, numbers, or words.

# Practice Page E

I have ___16___ dogs and cats.

How many of each could I have?

Keep track of your work. You can use pictures, numbers, or words.

# Practice Page F

I have __ 20 __ flowers and trees.

How many of each could I have?

Keep track of your work. You can use pictures, numbers, or words.

# Practice Page G

What are the missing numbers? Write them on the chart.

| 1 | 2 | 3 | 4 |  | 6 | 7 |  |  |  |
|---|---|---|---|---|---|---|---|---|---|
| 11 |  | 13 |  | 15 |  |  | 18 | 19 | 20 |
| 21 | 22 |  | 24 |  | 26 |  | 28 |  | 30 |
|  |  | 33 |  | 35 |  | 37 | 38 | 39 | 40 |
| 41 |  | 43 | 44 | 45 | 46 | 47 | 48 | 49 |  |
| 51 | 52 |  | 54 | 55 | 56 |  | 58 | 59 | 60 |
|  | 62 | 63 | 64 |  | 66 | 67 | 68 |  | 70 |
|  | 72 | 73 | 74 |  | 76 | 77 | 78 | 79 |  |
| 81 |  | 83 |  | 85 | 86 |  | 88 | 89 |  |
| 91 | 92 |  | 94 | 95 |  | 97 | 98 |  | 100 |

*Practice Page*
*Building Number Sense*

# Practice Page H

What are the missing numbers? Write them on the chart.

| 1 | 2 | 3 | | | | | | | |
|---|---|---|---|---|---|---|---|---|---|
| | | | 14 | 15 | 16 | | 18 | | |
| | 22 | | | | 26 | | 28 | 29 | 30 |
| 31 | 32 | 33 | 34 | 35 | 36 | 37 | 38 | 39 | 40 |
| | | | | | | | | | |
| 51 | 52 | | 54 | 55 | 56 | | 58 | 59 | 60 |
| 61 | | 63 | | 65 | | 67 | | 69 | |
| 71 | 72 | | 74 | | 76 | | 78 | 79 | 80 |
| | 82 | 83 | | 85 | 86 | | 88 | 89 | 90 |
| 91 | 92 | | 94 | | | | 98 | | |

*Practice Page*
*Building Number Sense*

# Practice Page I

What are the missing numbers? Write them on the chart.

| | | | | | | | |
|---|---|---|---|---|---|---|---|
| | | | | | | | |
| | | | | | | | |
| | 23 | | 25 | 26 | 27 | 28 | 29 |
| | 32 | | 34 | | 36 | | 38 |
| | 43 | | 45 | 46 | 47 | | 49 |
| 51 | | 54 | | | 57 | 58 | 59 |
| | 62 | 63 | | 65 | | 67 | 68 |
| | | | | | | | 80 |
| | 82 | | 84 | 85 | | 87 | |
| | 93 | | | 96 | 97 | | 99 |

*Practice Page*
*Building Number Sense*